For my mother.

And for the sixty million refugees and internally displaced people who are out there somewhere in the world today, risking their lives to reach safety.

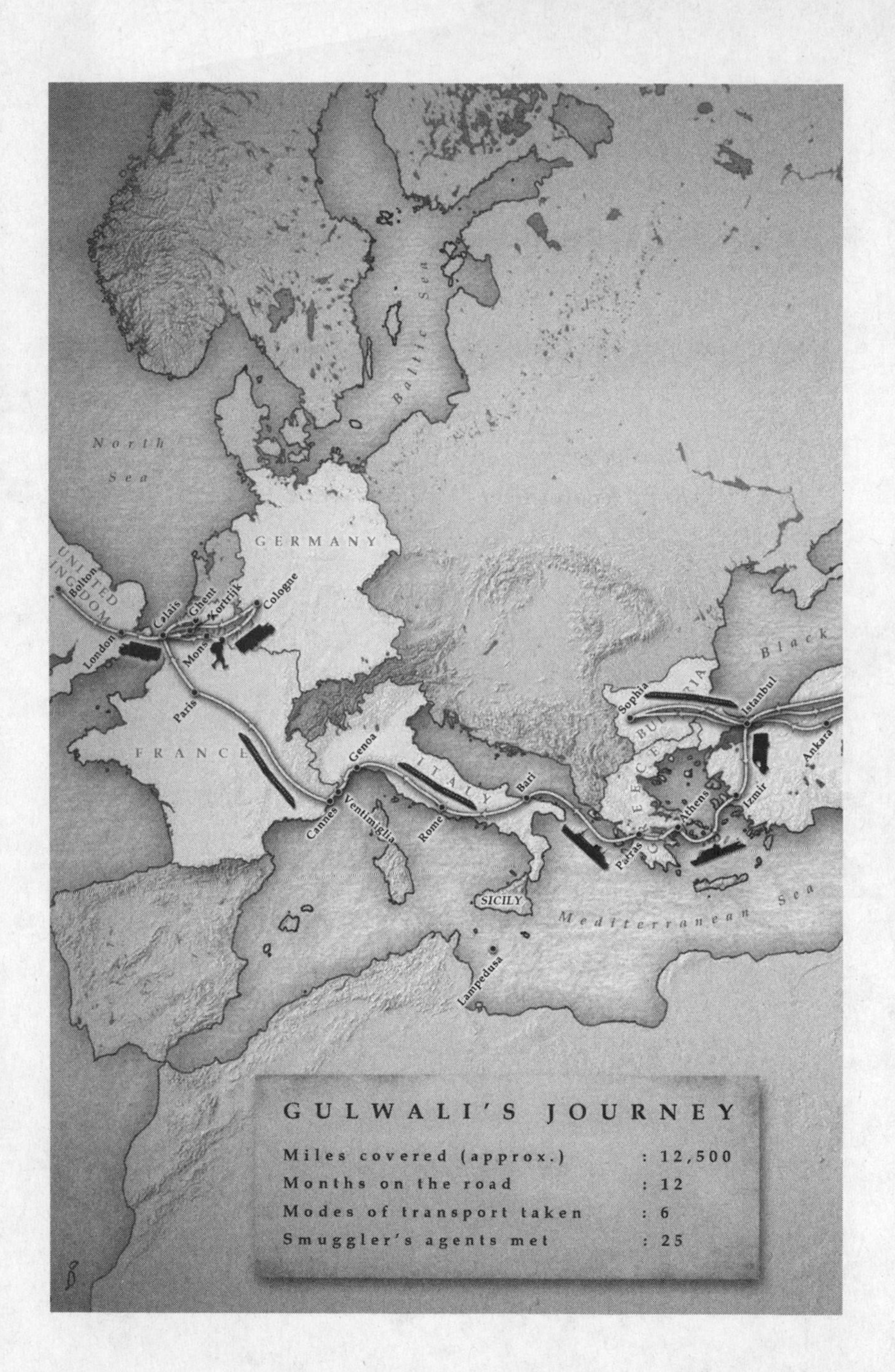
North Sea
Baltic Sea
GERMANY
UNITED KINGDOM
Bolton
London
Calais
Ghent
Kortrijk
Cologne
Mons
Paris
FRANCE
ITALY
Genoa
Cannes
Ventimiglia
Rome
Bari
SICILY
Lampedusa
Patras
Athens
Izmir
Sophia
Istanbul
Ankara
Black
Mediterranean Sea
GULWALI'S JOURNEY
Miles covered (approx.) : 12,500
Months on the road : 12
Modes of transport taken : 6
Smuggler's agents met : 25

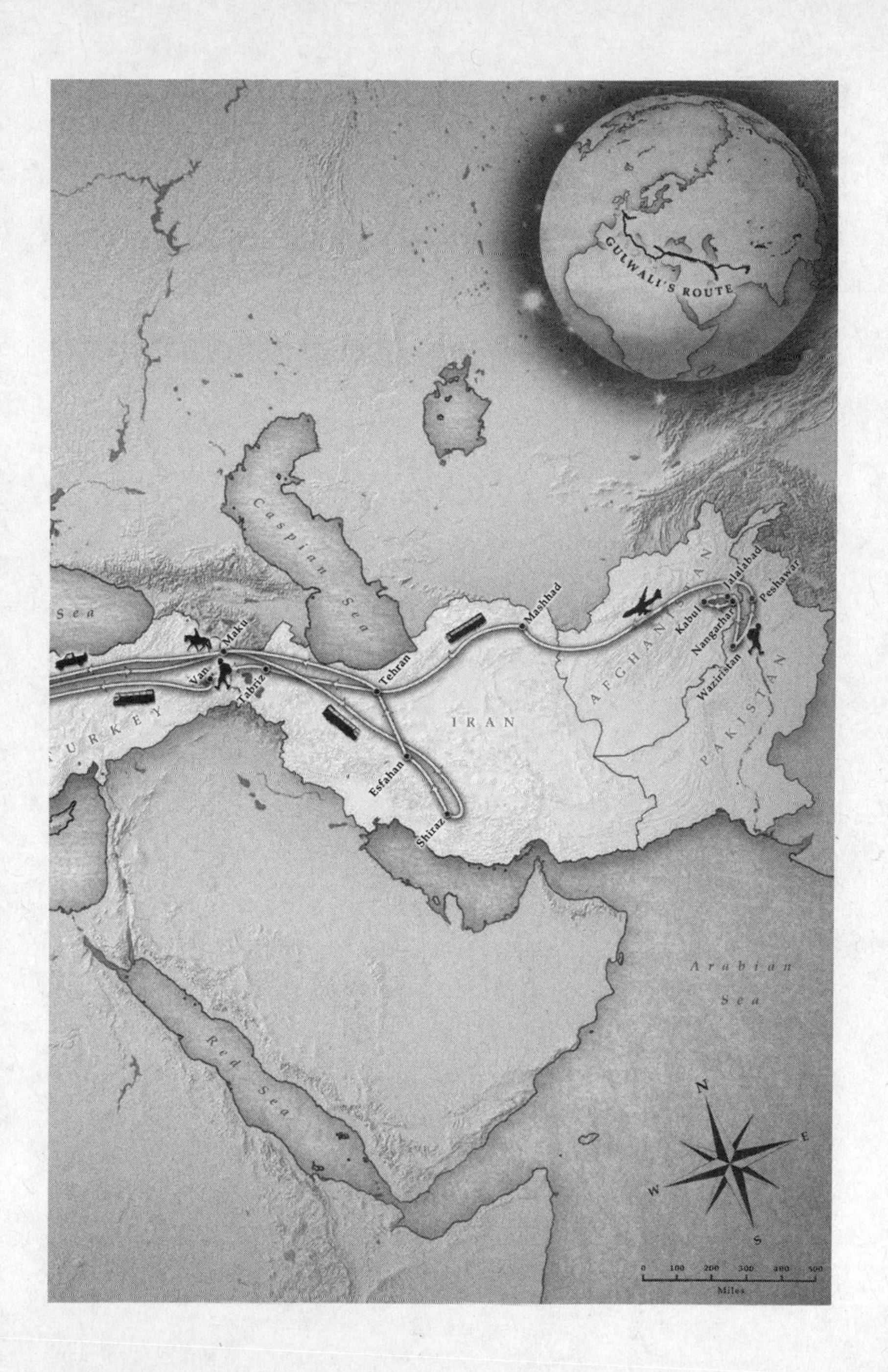

GULWALI'S ROUTE
Caspian Sea
Sea
Maku
Van
Tabriz
Tehran
Mashhad
Esfahan
Shiraz
TURKEY
IRAN
AFGHANISTAN
PAKISTAN
Kabul
Jalalabad
Peshawar
Nangarhar
Waziristan
Red Sea
Arabian Sea
N
E
S
W
0 100 200 300 400 500
Miles

PROLOGUE

BEFORE I DIED, I CONTEMPLATED HOW DROWNING WOULD feel.

It was clear to me now; this was how I would go: away from my mother's warmth, my father's strength, and my family's love. The white waves were going to devour me, swallow me whole in their terrifying jaws, and cast my young body aside to drift down into the cold, black depths.

"*Morya, Morya,*" I screamed, imploring my mother to come and snatch up her twelve-year-old son and lift him to safety.

The journey was supposed to be the beginning of my life, not the end of it.

I HAVE HEARD SOMEWHERE THAT DROWNING IS A PEACEFUL death. Whoever said that hasn't watched grown men soil themselves with fear aboard an overcrowded, broken-down boat in the middle of a raging Mediterranean storm.

We'd already eaten what little food and water the captain had on the boat. That had been more than a day ago. Now, fear, nausea, and human filth were the only things in abundance. Hope had sunk sometime during the endless night, dragging courage down with it. Despair filled my pockets like stones.

When we first set sail from Turkey, the white-haired Kurdish smuggler promised us we would reach Greece in a couple of hours. The man worked for one of the powerful, national-level agents, the shadowy businessmen who own and control the trade in the flow of desperate refugees moving through their countries. Money exchanges hands, and deals are struck through a series of regional agents and local middlemen. A powerful agent might have in his employ several junior agents and hundreds of local-level smugglers, drivers, and guides, dealing with hundreds or even thousands of migrants and refugees at any time, and in several different countries at once.

Yet, despite the Kurdish man's promises, it had been two days since we had set sail and we were still at sea.

On the morning of the second day, far out to sea away from any horizons, the captain changed the boat's flag from Turkish to Greek. This should have been a good sign, but something felt wrong. If we were in Greek waters, why hadn't we docked yet? Everyone guessed that something had gone awry and the majority of the men, many of whom were locked below in the hull, began to panic. These were the men who had been first to board, the ones who had shoved weaker men aside so that they might be guaranteed a place on the boat, a fair-weather tourist boat. As they boarded, the captain and his teenage crewman had instructed them to go below. How could they have known that they would then be locked behind a metal door? They hadn't expected to be trapped in a floating coffin, and they spent the night screaming, desperate to get out. I thanked the Creator I wasn't in there with them.

I was one of the last to get on and I was worried that I wouldn't get a space. By the time I was aboard, the hull was already full and I was placed on the open deck—a lucky stroke of fate. As the only child on the boat, my chances of survival weren't great even at the best of times, but at least being on the deck gave me a fighting chance.

There was no toilet anywhere on board. Men had soiled their clothes; others urinated into empty water bottles—some even saving their bright-yellow liquid to drink. Desperation can be a great motivator. A foul mix of sea water, urine, and feces lapped constantly at our feet and, even in the open air, the stench burned my nose. My bottom ached from the hard wooden bench that ran around the edge of the deck. We were wedged so tightly together, like sardines in a tin. It was impossible to snatch more than a couple of minutes' sleep at a time.

Hamid, a youth in his early twenties I had met just six days earlier, was sitting next to me. We rested our heads on each other's shoulders. My only other friend, Mehran, was one of the unfortunates trapped below deck. During the nights I heard him screaming in terror: "Allah, please help us. Allah."

The only reprieve came on the second night, when the captain allowed me and Hamid to go onto the roof of the boat. I don't know why I was chosen—maybe he felt a bit sorry for me because I was a small boy traveling alone. Big waves rocked the boat incessantly, but being high up felt safer, somehow. It was such a relief to get fresh air and to be able to stretch my arms and legs, but at the same time I was terrifyingly conscious that even the slightest wrong movement could see me toppling over the side and into the waves. I had no idea how to swim: if I fell in, I'd be dead. I didn't expect anyone would jump in to save me. Although we were all in this misery together, the rules dictated it was every man for himself.

By dawn of our third day at sea, the captain had become extremely agitated, constantly shouting into his radio in Turkish. I suppose he knew we couldn't stay out there for much longer without food or water. And if caught with a boatful of refugees, he'd be arrested.

I overheard a couple of the passengers, both Afghans like me, discussing whether it made sense to take control of the boat.

"Let's attack him and tie him up," said one.

His friend shook his head. "You fool. Who would get us into Greece if we did?"

The second man was right.

Like it or not, we were at the captain's mercy. His, and the sea's.

I felt delirious from lack of food and fresh water, and I started to hallucinate. My throat was so parched with thirst I was unable to breathe through my dry mouth. Maybe it was a way of counteracting the fear, but my mind started playing tricks. I kept thinking how nice it would be in Greece—just to wash my body, and not stink of piss and vomit. It sounds so stupid but I couldn't stop fantasizing about new clothes and how good they would feel on clean skin.

I think I was too focused on trying to stay alive to think much about the family I had left behind. My mother had paid people smugglers to get my brother Hazrat and me out of Afghanistan and toward what she hoped was safety in unknown lands. Instead, we'd both been thrust into separate hells.

It helped to try to focus on my mother's steely determination and imagine her voice urging me not to give up: "Be safe, and do not come back." They had been her last words to me and my brother before she had sent us away to find sanctuary in strange lands. She wanted to save our lives, to help us escape from men who had wanted us dead.

But so many times I wished she hadn't.

Sometime in the afternoon of the third day, the engine started to choke and splutter, then it cut out completely. The captain pretended that everything was okay, but, as time wore on, he grew even angrier, sweating and swearing as he tried to restart the ancient diesel motor. Eventually, he got on to his radio again and started shouting at someone, this time in a language I didn't recognize.

Finally, after one particularly heated conversation, he asked a Turkish speaker to translate.

"They are sending a new boat to get you," the translator announced. "Don't worry."

The captain smiled around at us all, displaying black, decaying teeth, but the look in his eyes gave the truth away, filling me with intense dread.

Not all of us were going to survive, of this I was certain. I felt rage swell inside me at the slippery lies that had come so easily from him.

My fears were confirmed when the weather worsened. Curling tails of wind whipped the waves into a frenzy, wailing like demonic beasts.

"*Morya, Morya*. I want *Morya*." Again, I cried out for my mother, far away in Afghanistan. I was a lost little boy, about to meet his death in a cold, foreign sea.

Before getting on this boat, I had never even seen the sea before; the only knowledge I had had of it was from pictures in school textbooks. The reality was beyond the wildest reaches of my imagination. For me, those waves were truly the entrance to the gates of hell.

I managed to get off the deck and onto a higher position—on the roof of the wheelhouse. The move gave me air and space, but now each rushing wave swung me back and forth like a rag doll. My skinny fingers gripped the railings, my knuckles white and bloodless.

After a couple of hours of this, the boat began taking in water. Everyone started screaming, the people trapped below frantically pummeling at the locked door with fists and shoes. "We are going to drown," they screamed. "Let us out. For God's sake, let us out. We will die here."

The captain waved a pistol and fired in the air, but no one paid him any attention. It seemed sure that the boat would overturn.

For a brief, strange moment I was calm, resigned: "So, Gulwali, this is how you will die." I imagined it—drowning—in explicit detail: the clean coolness of the water as darkness closed overhead, my life starting to flash before my eyes: my grandparents' wise, wizened faces; me at four years of age tending sheep by a mountain brook; walking proudly beside my father through the bazaar, him with his doctor's microscope tucked underneath his arm; sheltering from the baking sun under the grapevines with my brothers; the scent of hot steam as I helped to iron the clothes in my family's tailor shop; my mother's humming as she swept the yard.

No.

I wasn't giving up.

I had been traveling for eight months now. In that time, any childhood innocence had long since left me. I had suffered unspeakable indignities and dangers; watched men get beaten to a pulp; jumped from a speeding train; been left to suffocate for days on end in boiling-hot trucks; trekked over treacherous, mountainous border crossings; been twice imprisoned; and been shot at by border guards, their bullets whizzing over my head. There had rarely been a day when I hadn't witnessed man's inhumanity to man.

But, if I'd made it this far, I could make it now. A survival instinct deep within me spurred me on. I didn't want to die, not here, not like this, not gasping and choking for breath in the cold depths of the sea. How would anyone find my body?

My mother's face flashed before me again. "It's not safe for you here, Gulwali. I'm sending you away for your own safety."

How would she feel if she could see me now? Would she ever know what had happened to me?

That thought was enough to give me strength. I knew the captain had lied to us again—there was no other boat coming to get us, and this one was sinking fast. There was no way I was going to follow his orders to stay down and hide.

I searched in my bag and pulled out the red shirt I'd managed to buy in Istanbul, the one I was saving to wear as a celebration for getting to Greece. I started waving and screaming: "Help, help. Somebody help us."

I hadn't realized it, but the captain was behind me. As I turned, he kicked me full in the face, sending me tumbling down to the deck and almost over the side. Dazed and in agony, I clung on to the railing for dear life. The boat rocked back and forth but still I held my hand as high as I could, waving my shirt. The captain came for me again. I think he may have intended to push me overboard but by then others had followed my lead and had started screaming for help too, waving whatever they could to attract attention.

The boat gave a heavy belch and the bow dropped deeply into the water. Everyone screamed again and tried to move to the stern; I was still dazed from the captain's kick so could only try to protect myself from the stampeding legs.

The boat was finished, it was obvious. With a sickening wheeze, the stern settled heavily in the water too.

Now, we truly were sinking.

I closed my eyes and began to pray.

CHAPTER 1

"I FOUND YOU IN A BOX FLOATING DOWN THE RIVER."

I eyed my grandmother suspiciously.

Her deep brown eyes danced mischievously, set within a face that was deeply lined and etched by a lifetime's toil in the harsh Afghan sun.

I was four years old, and had just asked the classic question of where I'd come from. "You are joking with me, *Zhoola Abhai*."

Calling her "Old Mother" always made her smile.

"Why would an old woman lie? I found you in the river, and I made you mine."

With that, she let out a toothless chuckle and wrapped me in her strong arms. I was my grandparents' second grandchild, born one year after Hazrat, but I felt like I was their favorite, with a very special place in their hearts.

MY FAMILY IS FROM THE PASHTUN TRIBE, WHICH IS KNOWN for both its loyalty and fierceness. Home was the eastern Afghan province of Nangarhar, the most populated province in Afghanistan, and also a place of vast deserts and towering mountains. It is also a very traditional place, where, even today, local power structures continue to run along feudal and tribal lines.

I was born in 1994, a year before the Taliban government took control of Afghanistan. For many Afghans, and particularly for my family, the rise of the ultra-conservative Taliban was a good thing. The Taliban were seen as a stabilizing force, one that brought peace and security to a country that for more than fifteen years had suffered unimaginable hardships and endless violence—during first the Russian invasion, then, later, a brutal civil war.

For much of their marriage, my grandparents lived in a refugee camp in the northwestern Pakistani city of Peshawar. The refugee camp was also where my parents met and were married. By the time I was born, Afghanistan was not at war, and was relatively stable under Taliban rule.

My earliest memory is of being four years old and running with my grandfather's sheep high in the mountains. Grandfather, or *Zoor Aba* ("Old Father"), as I called him in my native language of Pashtu, was a nomadic farmer and shepherd. He was a short man, made taller by the traditional gray turban he always wore. His hazel-flecked green eyes shone with a vital energy that belied his years.

Each spring, he walked his flock of thickly fleeced sheep and spiral-horned cattle to the farthest reaches of the mountains in search of fresh and fertile pasture. My grandparents' home, a traditional tent made from wooden poles and embroidered cloth, traveled with them. Two donkeys carried the tent on their backs, along with the drums of cooking oil, sacks of rice, and the flour my grandmother needed to bake naan. I used to watch, transfixed, as my grandmother spread and kneaded sticky dough along a flat rock before baking it over the embers of an open fire. She cooked on a single metal pan, which hung from chains slung over some branches balanced over the fire. I loved helping her gather armfuls of wild nettles, which she boiled to make a delicious, delicately scented soup. I don't know how she did it, but everything she created in that pan tasted of pure heaven to a constantly hungry little boy like me.

Every year, as the leaves began to turn into autumn's colors, my grandparents would head back down to lower ground, making sure to return to civilization before the harsh snows of winter descended and trapped them on the mountains' slopes. There they joined the rest of their family, their six children and assorted grandchildren, in the rambling, stone-built structure that housed our entire extended family. Though simple, our home was lovely, perched above a clear, flowing river.

Grandfather loved his family with a fierce passion, and laughter came easily to both him and my grandmother. I don't think I ever saw him angry. One time I accidentally almost took his eye out with a catapult. Blood streamed down his cheek from where my badly aimed rock had cut it. It must have really hurt but he didn't chastise me. Instead, with characteristic humor, he managed to make a joke of it: "Good shot, Gulwali."

My grandmother was sturdily built and bigger than my grandfather. She was definitely the boss, but I could see they adored each other. Love isn't something people really discuss in Afghanistan. Families arrange marriage matches according to social structure, tribal structure, or even to facilitate business deals; no one expects or even wants to be in love. You just do as your parents demand and make a marriage work the best you can—you have to, because divorce is forbidden for women.

It was explained to me once—by my grandfather—that a woman is too flighty and unsure of her own mind to understand the consequences of leaving her marriage. Besides, who would look after her if she did? Men do have the right to divorce their wives, but it is still frowned upon in Afghanistan. I knew only of one woman whose husband had divorced her. She'd been taken in by her brother, but she remained a great shame to her family. She had been lucky that he had accepted her and hadn't turned her away onto the streets.

My grandparents would never have dreamed of breaking up, even if they could have done. They had married when she was fifteen and he eighteen, meeting for the first time on the day of their wedding, as is still often the norm. But anyone could see that their years together had given the pair a special bond.

Because I was my grandparents' shadow, when I was three my parents agreed to let me go to the mountains with them. For the next three and a half years, I shared my grandparents' nomadic lifestyle along with their youngest daughter, my auntie Khosala ("happy"), who was twelve years my senior and like a big sister to me. At night, we slept soundly beneath a vast, star-filled mountain sky, safely tucked up inside the tent nestled between the pair of them.

By the time I was five, I was already a skilled shepherd, able to shear off a fleece all on my own. I recognized every animal individually and loved how they knew the sound of my whistle. I particularly enjoyed watching my grandfather's two sheepdogs working. One was a large, thick-headed beast called Totie and the other a small, wiry terrier-type dog we named Tandar. They would run rings around the flock, corralling them into order. And when the local vet, a man who traversed the farthest reaches of the mountains to service his clients, came to treat the sheep, I remember thinking how brilliant it might be to be a vet myself when I grew up. I was fascinated by him and the various implements he used.

It was about as wonderfully simple and rural a life as you can possibly imagine.

In winter, I would proudly come back down into town with Grandfather by my side. We carried with us precious bounty from the mountains: wild fruits, honey, and *koch*—a type of thick, unpasteurized butter that we would spread thickly on freshly baked naan for breakfast. And Grandfather would always take me to the bustling bazaar, where he would trade his wares for supplies of rice or a new farming implement. Everything was plentiful.

Returning to the family home meant I got to see my parents and siblings. Although I loved being with the sheep, I did miss my parents. And, of course, they'd missed me too, so I was very spoiled when I came home.

As is the custom within our tribe, my parents were distantly related: my mother was my paternal grandfather's niece, his sister's daughter. My mother was fifteen and my father twenty when they married in the refugee camp to which my grandparents had fled after the 1979 Russian invasion of Afghanistan. During the fifteen or so long years of occupation and the civil war that followed, it is estimated that some three million people—a third of Afghanistan's population—died, while a similar number fled the country as refugees.

In the midst of all this chaos, my grandfather had somehow scrimped and struggled enough to ensure that my father, his eldest son, became the first man in the family to receive a higher education by studying to become a doctor. My father's profession was a huge source of pride for my family, and demonstrated that my grandparents, whose sacrifices ensured my father's success, remained the moral heart of our family.

My father's brothers, my two middle uncles, were also successes. They were both tailors who ran a large and profitable workshop in the bazaar. The fourth and youngest uncle, Lala, wasn't around as often. He had a senior role within the Taliban. He used to visit us, bringing Taliban soldiers with him. I thought he was cool and exuded power. I knew he was an important man but didn't really understand why or exactly what it was he did.

My mother's parents had stayed in Pakistan, so at that time I didn't know them very well. My mother was one of twelve daughters. Her father was a very educated man—a mullah—and he had educated his daughters, something that was quite unusual among Pashtuns in those days. My mother was the only woman in our entire household who could read.

I think my parents were happy together: they certainly seemed it. But in Afghanistan a child knows better than to discuss or ask these things. There are certain boundaries you do not cross. Once, though, I did ask my grandmother if she liked my grandfather. She just laughed and replied: "I think he was the one who liked me." As innocent as it sounds, in our conservative community that was quite a risky thing to say, even for our beloved grandmother.

My parents had three boys by then: me; Hazrat; and Noor, who was a year younger than me. Hazrat and Noor were very close, and they used to pick on me. I was jealous of their little world of private jokes and unspoken communication. I think I was a bit of a loner, possibly because I was used to the solitude of the shepherding life, which I continued until I was six, when my life changed completely.

WORRIED THAT MY GRANDPARENTS WERE GETTING TOO OLD for the nomadic life, my father and my three uncles ordered their parents back home. My father and uncles wanted them to stay closer to the family so that they could be better looked after. There was also the matter of family honor: my father's profession as a doctor meant he was a highly respected man in our strictly conservative community. It didn't look good within the rigid mores of our tribal society that *his* father lived like a poor *kochi,* or nomad. Such was my father's standing, in fact, that my brothers and I were rarely referred to by outsiders by our names: we were known as "the sons of the doctor." Even Grandfather was known as "father of the doctor."

Both my grandparents loved their life, so were deeply resistant to the idea, but ultimately gave in. Grandfather sold his entire flock of sheep—more than two hundred strong—for a combination of cash and a shiny new red tractor. The whole extended family—my parents, my grandparents, my two married uncles, Auntie Khosala, and me and my brothers—then moved to a new house in the district of Hisarak. This house was another single-story building, made of mud and thatch, with lots of rooms running off a central, communal kitchen. Each night we ate together sitting on the floor, a bounty of food—usually rice, meat, naan, and spinach—spread out on a large tablecloth in the center of the room. It was a happy home, full of chatter and noise. I still loved the company of my grandparents and insisted I sleep in their room.

My mother, as the senior wife of the family, managed the running of the home, while my uncle's wives, junior in both age and position, did the majority of the cooking and cleaning. Like most Afghan women, her outer nature was steely and unemotional, and it was no wonder. Most women work from dawn until dusk doing housework. Washing is done by hand, while wood and fresh water must be collected daily. There is always bread to be baked and hot tea to be freshly brewed before husbands and children awake in the morning. It's also a land where two out of five Afghan children die before their fifth birthday, so it is easier not to show too much love to children. A year after Noor was born, for example, my mother gave birth to twin boys, who sadly both died within a few days.

But love is hard to hide when it's a part of your very being, as it was for my mother. Under her commanding exterior, her survival mask, I often saw her gentle side—the way she would fuss over me when mending a scrape or bruise, her worry when one of us was ill, her obvious pride when recalling our accomplishments to a visitor or one of my aunts. She had a very deep voice for a woman, but was tall and elegant, with a long nose and round, brown eyes. People said I looked like her. She smiled rarely, except for the discreet grins that flashed across her face when I was naughty or did something funny. She was tough because she had to be, but there was an unmistakable warmth to my mother, and I knew even as a young boy that her family meant everything to her.

Culturally, it was a great shame to allow women outside in case they were seen by other men, so my mother and aunts rarely left the house. On the rare occasions they did, they were completely covered by a burqa—as was the rule under the Taliban government. Inside, they wore long shawls to cover their hair. It would have been seen as very bad for anyone outside of the immediate family, even a male cousin, to see their heads uncovered, even in the house.

I was an extremely pious child, which I learned from my uncles Lala, Haji, and Thedak, who were similarly devout, much more so than my father. Following my uncles' example, I liked to enforce the rules of Islam. "The wrath of Allah will be upon you," I used to say to my aunties, all nicknamed Tindari, a term which means "uncle's wife." "Go and cover your head." If I wasn't playing with my brothers or cousins, I often sat with them in the kitchen, bossing them around and ordering them to bring me tea, even though they were working hard baking naan and cooking over the open fire. When my uncles were away, I would often refuse to let them walk to collect firewood, visit people, or attend family weddings. I saw this as protecting the family honor. I would make a big show of insisting on collecting the wood for them so they didn't have to: "Why do you need to go outside?" I would say. "You are the queens of this house." This was something I'd heard my uncles say many times. As was another saying in Pashtu: "*Khor yor ghor,*" which was the two places for a woman: home or grave. Sometimes I would wake my aunts in the middle of the night to bring food for a newly arrived guest, which happened regularly because my father's patients would turn up at our house at all times of the night and day.

I was a bratty child, and I enjoyed exerting my power over my aunts. I know they loved me, but I think I must have really got on their nerves at times. And if they complained about my behavior, my uncles would tell them to be quiet and to obey me. It was not the best way to keep a child's ego in check, but this was how it was done. In our conservative culture, males have all the power, even little boys.

The only time I got seriously told off for picking on my aunts was by my mother. One of my uncle's wives couldn't have children, and this was a source of consternation to the whole family. "If you don't get a baby soon, I will get my uncle a new wife," I rudely said to her one day. The poor woman cried. My mother was absolutely furious with me and made me go and apologize at once. My aunt hugged me and I remember realizing I'd said something really mean, even though I didn't understand the severity of it at the time. It took nine years for her to get pregnant, but she went on to have six daughters.

Another of my aunts, my father's sister, Meena, married a man who lived outside the district. This was a really big deal because it was the first time anyone in the family had married outside the tribe. Many people, including my uncles, were not happy about it.

Auntie Khosala was next to be married. She and I had a special bond because of all the time we'd spent together in the mountains, and I felt sad for her because although she couldn't read—she was naturally very smart—the man she married, her first cousin, was not only illiterate but obviously thick. But the match had been arranged when she was a baby: my grandmother had been pregnant at the same time as the wife of Grandfather's brother, and when the babies were born within days of each other, my grandfather and his brother had decreed that the two infants would marry when they were older. Within our culture these things are not said lightly; once said, they cannot be unsaid, and must go on to happen.

Aside from being a farmer's wife, my grandmother was also a traditional midwife. Pashtun men do not like to take their wives to a doctor—it's considered extremely shameful if anyone, especially a man, puts his hands on your wife. But the Taliban government had banned female doctors from practicing. In those circumstances, it was not surprising that since coming back from the mountains, my grandmother's skills had been in great demand. Some men would have forbidden her from her midwifery work, especially because it often meant being out on the streets at all hours. But Grandfather was very proud of her and liked to joke about the hundreds of babies she'd "given birth" to.

I often went with her to attend the births, but it was not something I enjoyed. My grandmother would leave me outside, where I helped look after the family's children or talked to the woman's husband. Sometimes, if the family was poor, their house only had one room, so I sat in the corner as the woman screamed and bled. Childbirth horrified me. The labor could go on for hours. I would sit quietly, childishly willing and wishing the woman would get on and push the stupid thing out so we could go home.

My grandmother knew I was squeamish about the whole thing so she teased me mercilessly: "Did you see all that blood, Gulwali? Did you look, you naughty child?"

"No, I did NOT."

"Ahhh, you lie to me." She'd give me a toothless grin and rub her fingers through my hair.

"Get off me. You were touching those women."

That only made her cackle more as she grabbed me in a bear hug, wiping her hands all over my head as I squealed with horror.

My family owned various shops in the bazaar, including the tailor's workshop run by my two middle uncles, and my father's doctor's surgery. Along the flat roofs ran a network of vines laden with fat grapes, which we sold commercially. We also had many fields out of town that grew wheat and different vegetables; at busy harvest times, we employed as many as one hundred men locally. And before the Taliban took over and banned it, my grandfather had also farmed opium and cannabis—something that was entirely usual for Afghan farmers.

In warm weather, the whole town would become burningly hot, and the brown sandy dust of the town's desert landscape got everywhere, stinging our eyes, blocking our noses. My mother and aunts would do everything they could to keep all of the dust outside the house, seemingly spending all day sweeping with a stiff broom, or banging rugs against the walls. Their efforts were futile. The sand collected in little piles in the corners of the rooms, on door frames, and behind chairs, and it covered window sills. And with the dusty sand came all manner of insects and bugs: dangerous scorpions, and the little black ants, which fascinated me. I loved to watch them mobilize into lines of activity and scurry toward the kitchen. Of course, keeping ants out of the rice stores was another lost battle. Whenever I heard my mother let out a long groan as she opened one of the earthenware containers that kept the grains cool, I chuckled because I knew how funny her face would be as she scowled at the infestation of ant invaders in our dinner-to-be.

After my father curtailed my grandparents' nomadic activities, I was enrolled in the same local school that my brothers attended. On my first day, my uncle literally had to drag me there: "I'm not going," I yelled all the way up to the classroom door. "I don't want to go." I wanted to go back to the mountains and run with the sheep, not sit stuck in a classroom. But I soon settled in and became a very studious, hard-working pupil.

The most fun we had at school was in winter, when we had to fix the roof of our classroom. The building had been badly damaged during the Russian invasion, so none of the rooms had ceilings. In the summer that wasn't a problem, but in winter each class was dispatched to cut down trees and help drag them back to make a temporary roof. I loved it.

In Afghanistan, children aren't coddled; they are expected to pitch in and work alongside the adults. Before mosque each morning there were duties in the fields: cutting hay for the cows, watering the crops, harvesting the vegetables, collecting firewood and jerry cans full of water from the nearby well. All the kids went, even the two newest additions to our family, my little sisters, Taja and Razia. As female education was banned by the Taliban government, during the day they helped my mother and aunts with the cooking and cleaning. In our culture, it's seen as important that girls learn these duties from a very young age, so that by the time they are married they know how to manage the running of a household. They also learned how to sew and embroider, as well as studied the Quran. After those chores, we went to the local mosque to pray, before getting ready for school.

We didn't really play outside on the street; the kids who played outside were seen as the bad kids. My parents thought it showed bad manners and a lazy attitude. So, after walking home from school, my brothers used to hang out in the grapevines, making sure nobody tried to help themselves.

I preferred spending time with the tailor uncles, Haji and Thedak. I was fascinated by the whirr of the sewing machines and the folds of the differently colored woven cottons and wool they used to make *shalwar kameez,* the traditional, loose-fitting tunic and trousers worn by Afghan men. At times of celebration, they were overrun with orders, and my uncles would be at the shop day and night sewing wildly to get everything finished. I used to help by cutting out the fabrics and dealing with the customers.

They were very proud of me when I devised my own system of numbering and organization, so that when people came to get their clothes I knew, among the hundreds of garments, where everything was. I took joy in the order and neatness of my system: to me, it was a little world of my own—a way of bringing a sense of order. A long day spent helping them checking off my lists and hanging the clothes ready for the next day was one of my biggest pleasures. By then, I knew I definitely wanted to become a tailor too. I was very enterprising—I even had my own little after-school stall outside their workshop, which sold threads and buttons.

The journey to and from school took us right past my father's surgery, so occasionally Hazrat, Noor, and I would pop in to say hello to him. I used to sweep the floor for him or go and fill a jug with fresh water from the modern pump near the district governor's house. I always felt sorry for him working so hard in his packed surgery, so I wanted to make sure he had fresh, cool water to drink.

In the early mornings, my father would do his rounds in the local hospital. Treating malaria, the mosquito-borne disease that blighted the local area, was his specialism. For the rest of the day, he ran his surgery, and people came from miles around for him to treat their ailments.

His surgery was full of strange-looking equipment. I recall a little wooden box that used to store his dentistry tools. I remember watching in horrified fascination as he pulled a patient's head back and used metal pliers to prize out a rotten tooth. Whether he was stitching a wound, healing a snake or scorpion bite, or setting a broken limb, my father always wore on his face a mask of studious concentration.

While I remained taken with his teeth-removing tool, my father preferred his microscope, a very rare thing in those days. It was too valuable to leave in the surgery so, at night, he used to carry it home under his arm, covered with a black cloth to keep it free from dust.

Sometimes, when the surgery was quiet, he used to sit outside in the sun on a wood-framed bench with a seat of knotted string, and chat with his friends or argue about politics. I used to love listening as he and his friends debated the ills of the world, while I sat quietly, cross-legged on the ground, taking it all in.

I got the sense that he was a good man, a generous one. He was immensely popular: he couldn't walk through the bazaar without people wanting to shake his hand or offer him gifts of fruit or sweets. He had two big books in his office detailing the people to whom he had loaned money, and he never turned anyone anyway if they needed medicine and couldn't pay. The ledgers were very big—clearly, a lot of people owed him money.

The most intimate times I spent with my father were during the night, when patients who had walked from outlying villages with a medical emergency banged on our gate, hoping to be seen by him. To treat them he would have to take them to the surgery where his bandages and medicines were stored. In those days, electricity was scarce and only came via a generator, something we limited to using three or four hours a day. As my father helped the patient, I would walk ahead carrying a lantern to help light our path through the darkness. Stepping through the pitch-black streets with the lamp held aloft, I felt of great consequence, making sure no one stepped into a puddle or twisted their ankle in a pothole and ensuring everyone's safe passage.

One of my most treasured memories was when my grandfather ran away with me. My father had told me off for reasons I can't remember—I must have done something very naughty indeed, because my father rarely chastised me. My grandfather scooped me up into his arms and set off into the mountains.

"You don't deserve this boy," he yelled to my father. "He's coming with me."

In reality, I was a convenient excuse for the wily old man to get away for a few days. He had wanted to go on a trip to search for lapis lazuli, the blue semi-precious stone that is one of Afghanistan's most precious natural resources. But he knew that there was no way his sons would approve, so engineering a fight with my father over me had given him a convenient way out.

He and I had a truly wonderful week playing truant and searching for the stones. When we came back, my childish misdemeanors had been long forgotten—Grandfather was the one in trouble this time.

It was to be my last happy memory before my whole world changed forever.

CHAPTER 2

EVEN BEFORE THE ARRIVAL OF THE TALIBAN, MY FAMILY was religiously and culturally conservative. We lived by *Pashtunwali*—the strict rules every Pashtun abides by. The codes primarily govern social etiquette, such as how to treat a guest. Courage is a big part of the Pashtun code too, as is loyalty and honoring your family and your women. No one writes it down for you—there's no book to learn it from; it's just a way of life that you are born into, which has remained unchanged for centuries. And, like all Pashtuns, I accepted it unquestioningly. It was—and still is—a source of cultural pride.

The Taliban government invoked certain elements of *Pashtunwali* to suit their purposes, such as the need to honor and protect women by keeping them indoors. The Taliban were very effective politicians at the local level. They saw it as their moral duty to cut crimes and keep the peace on the streets. Making people feel more secure is one of the reasons they had so much support from the conservative elements of Afghan society. My father told me that before the Taliban came to power, toward the end of the civil war, women and young girls were frequently raped if they went outside, while houses were robbed and children kidnapped and held for ransom. He said the country had been in ruins, but that the Taliban had returned order. Their position was simple: strong social controls and Sharia law were the true path to both peace and God. To me and the male members of my family, this made perfect sense, because it wasn't really so different from *Pashtunwali*. No one asked the women for their views, however—they didn't get to have a say in these kinds of political matters. But I think the whole family was in unity. We all understood our different roles within the household.

Still, I was aware that not everyone agreed with us. A few of my friends at school told me their parents feared the Taliban. They said the Taliban were taking our country in the wrong direction, preventing development. They also said that they abused women.

When I heard this, it made me angry. My uncle Lala was a senior member of the Taliban. How could they possibly be bad? And yet their rules certainly were strict—men's beards had to be a certain length and men had to wear traditional, baggy clothes at all times, while women had to wear the burqa so their faces couldn't be seen, and soft-soled shoes so they couldn't be heard. Under the Taliban, everyone was required to pray five fixed times of the day. Praying five times a day is an Islamic requirement, but people generally took a relaxed attitude to timings or locations of where they prayed. Under the Taliban regime, the moment the mosque loudspeakers blasted out the hauntingly beautiful call to prayer, everyone in the whole town would stop whatever they were doing and go straight to mosque. If you were seen on the streets or caught working during fixed prayer times, you were arrested. The Taliban had a special team, called the Vice and Virtue Committee, who made sure everyone obeyed. They wore black turbans and patrolled the streets in four-wheel drives, pulling over and leaping out of the vehicle if they saw something amiss.

One Monday afternoon, we were in Haji and Thedak's tailor shop when we heard a big commotion outside. A man stuck his head through the shop door, his face ablaze with excitement: "Come on. Twenty lashes."

My uncles quickly locked up the shop and urged me to follow them to the bazaar. A large crowd had already gathered at the scene: two men in black turbans stood over a kneeling man, who was naked from the waist down.

"Speak. Admit your crime," said one of the turbans.

The man said nothing.

"Speak." The turban smacked the back of the man's head.

"I—"

"Louder."

The man mumbled something, but it was impossible to hear it over the hum of the crowd.

The turban turned to the crowd with a smirk. "This man did not attend prayer on Friday. He was found inside his home."

The crowd jeered at this seemingly obscene crime.

"Do you have an excuse?" the turban asked.

This time the man looked up and spoke loudly. "My wife was sick. I couldn't leave her. I thought she was dying."

At this, both turbans roared with laughter. The crowd followed suit.

This time, the second turban spoke: "The Islamic Emirates of Afghanistan gives the sentence of twenty lashes with justice to this man for his crime." His voice rang out as clear as a bell. "May this sentence be an example for anyone else thinking of committing such a crime. Anyone who goes beyond the limits of God's law and against His rules, let it be known this will be their fate."

Then he lifted a thin reed high into the air.

"By Allah the most merciful, you have been sentenced to a fitting punishment for your crime."

As he brought the reed down on the man's back with a loud swish, the crowd roared and cheered.

The turban hit the man again and again and again.

By the time it was over, their victim had crumpled into a heap, his back a bloody mess. I don't know if anyone helped him or if they took him to prison after that, because my two uncles said it was time to get back to the shop.

Five minutes later, the sewing machines whirred into life and I was back to hanging clothes as if nothing had happened.

Many times after that, I saw people get lashed or beaten. I once saw an old man get beaten with a wire cable because his beard was too short: the black turbans would grab a beard with their hands to measure it—if the person measuring it had big hands and you had a wispy beard, then it was unlucky for you. Other times they gave out smaller punishments, such as painting someone's face black and marching them around, encouraging people to abuse them.

I suppose it sounds shocking, but this was all I knew. I had been taught by Uncle Lala that the Taliban only punished someone when they broke the law, so in my child's mind if someone was suffering they must have deserved it.

But one incident will always haunt me.

Grandfather and I were with a friend of his on our way to visit some relatives in a different district. We had stopped for *chai* in a small town. As we sat outside the tearoom, we watched several men begin running in one direction. My grandfather stopped a man to ask what was happening.

"Justice, my friend. Justice."

Curious to see what was happening, we followed. At the edge of the village, scores of men had gathered in a large clearing. Some of them were shouting, "*Allah akbhar,* praise be to God," while others were whooping with joy.

At first I thought it must be some kind of sporting event—maybe a dog- or cockfight. I strained my head to see what they were looking at. When I couldn't see past the men gathered in a circle, my grandfather's friend lifted me onto his shoulders.

I saw a woman. She was covered in a burqa with a black blindfold tied around her eyes, her hands behind her back, cinched tightly with a rope.

A man next to her—I guessed a Taliban official from the Justice Department—raised his hand. The crowd went silent. "This woman is a dirty woman. She is an adulteress, a whore."

The crowd roared insults at her.

"The Islamic Emirates of Afghanistan gives the sentence of death to this woman for her crime. May this sentence be an example. She has been condemned to death by stoning. Let it begin."

At that, he placed his hand on the woman's shoulder, almost tenderly, and walked her toward a dug-out area of earth. He took her arm and helped her step down into it until her waist was even with the dirt line. She didn't try to resist.

It was then that I noticed the rocks, sitting in a pile to the front of the crowd.

The Talib walked to the pile and picked up a large stone. He waved it above his head like a trophy. "By Allah the most merciful, you have been sentenced to a fitting punishment for your crime."

He walked back toward her, until he was a couple of feet away.

Then he threw the stone hard. It smashed against the woman's head.

The crowd went wild. All at once men began picking up the rocks and throwing them at the woman. At first, she didn't seem to move, even as they hit her, but then one little rock caught the back of her head sharply, and she began to try to wriggle her way out of the pit. But as she did so, another hail of stones and rocks caught her and she fell backward. After that, it was a bit of a blur for me, just a hail of rocks and a lot of shouting.

The whole thing only took a couple of minutes.

I don't think I really understood what I was seeing. The noise, the excitement of the crowd, the calm silence of my grandfather and his friend. I didn't enjoy it, but I didn't cry either. As I sat on the shoulders of Grandpa's friend, I looked around and saw other boys my age similarly perched on the shoulders of grown men. Their expressions were blank, but their eyes were wild with confusion, mirroring my own.

With hindsight, I can see what barbarity this was. But it has to be put into context. The country was recovering from a war where brutality had known no bounds. Millions of people had been killed, thousands of women raped and children slaughtered. There had been a complete breakdown of law and order with no national governance. Their mantra was this: "If you don't accept peace or our rules, we will force them on you."

My grandfather called life under the Taliban: "the best of times we are living in." Given that he'd lived most of his life in the insanity of war, I don't think it is too surprising that he and many other Afghans felt this way.

But, as is ever the case in the geopolitics of my country, we were about to get out of the frying pan and into the fire.

CHAPTER 3

I WAS SEVEN YEARS OLD.

I remember running between the kitchen and the guesthouse, my mother handing me pots of tea to take to the assorted tribal elders and Taliban fighters who had suddenly amassed in our home. The air was thick with tension, heavy with discussion.

"We didn't have anything to do with it," one man said, "so why should we bow down to the imperialists?"

A Saudi man called bin Laden, who Uncle Lala told me was a great freedom fighter, had attacked the United States. TV was banned under the Taliban so we didn't see any images of it, but the radio had been broadcasting nonstop news about it. Thousands of people had died in the attacks. They said people had been jumping from windows to escape. I had been told that Americans were infidels who didn't follow Islam, but as I listened to these stories, I felt sad at the news and was thinking about the families of the people who had died. Given my family history of living in refugee camps, I think I had an innate understanding that, wherever they were from, all people suffered in war and conflict.

Now, just days after the attack, the United States blamed Afghanistan for bin Laden's attack. This was because the Taliban refused to bow to American pressure and hand over bin Laden, who was said to be sheltering in Tora Bora, a place only a couple of hours' drive away from where we lived.

"Bin Laden is our guest," one man insisted. "*Pashtunwali* must prevail."

"Our great country will never cede to this pressure."

The United States had threatened to attack if they didn't hand him over, but the Taliban refused because, under the rules of *Pashtunwali,* he was our guest, and a guest is under the protection of the host.

There was a swell of patriotism. No one really believed the United States' threats, mostly because everyone knew the Taliban had had nothing to do with the attack themselves, and we assumed the Americans accepted that too.

But, it turned out, this wasn't a view shared or understood by the world. Within days, the United States, supported by forty-eight other countries, invaded my country, sending the Taliban into retreat.

WHEN THE INVASION STARTED, WE COULD SEE U.S. FIGHTER jets overhead. For a day or so they only circled the skies, not doing anything. Taliban troops were everywhere, and lots of local people were volunteering to fight with them.

My father expected the worst. He busied himself trying to amass extra medical supplies. He ordered my mother to pack bags and food and take the children to the shelter of a nearby bunker. There were several concrete bunkers on the outskirts of town, a sad, if practical, legacy of the Afghan civil war.

The following day, the United States started bombing Tora Bora. The mountain was far enough away, so we weren't in any real danger, but within days, bombs started raining from the sky directly overhead. We huddled in the bomb shelter with other families.

It was terrifying.

It was never clear exactly where the planes would come from. We would hear the screech of Taliban anti-aircraft missiles attacking them as they approached, then there would be the sound of deafening, terrifying explosions. The whole ground used to shake with the force of the bombing. After each bombardment, we would come outside for air and look up at the trails of smoke the B–52 bombers left behind in the sky.

Pretty much every type of bomb except nuclear rained down on my country. Even today, children are born with diseases and deformities caused by the toxic effects of that time, something that still makes me very angry.

Uncle Lala was the regional commander for the Taliban, leading the fight against the invaders. My father stayed in the hospital, treating the wounded. My grandfather insisted on staying in the house to protect against looters. I was proud of them all, for different reasons, but I was most worried about my father. I feared the bombs might hit the hospital.

I longed to run to the solace of my mountain paradise, but it wasn't safe. Fighting echoed across the hills. It seemed the whole of Afghanistan was under attack, and for days we couldn't sleep. I thought it was only a matter of time before the bombs would destroy our bunker.

It was only later that I learned that the area around our home had been the last front line; Uncle Lala had been one of the last local commanders of the Taliban to hold his position against the U.S.-led coalition forces. We were told that his courage allowed busloads of fighters to escape Kabul into neighboring Pakistan; in my childish eyes, this made him something of a hero. Soon after that final battle, Lala fled the country, and we didn't know where he had gone. I feared he'd been caught and killed, but we had no way of knowing.

Life began then under occupation. Suddenly, U.S. troops were everywhere, convoys of armored vehicles speeding down the road. The first time I saw Western troops on the ground, I was so scared of their big guns. They looked like something from another planet. From a safe distance, my brothers and I would throw stones at them.

After a while, seeing them became normal.

THE PRESSURE AND POLITICAL TURMOIL IMPACTED ALL OF US. My father decided to move the family to a neighboring district, his place of birth, where things were slightly safer.

Not long after the U.S. troops landed, the aid workers arrived—white Land Cruisers emblazoned with the blue UN sign were everywhere. They started to rebuild clinics and schools. My father conceded this was a good thing, but he still disapproved of "foreigners" building our country's infrastructure instead of our own government.

Girls, who had been banned from education under the Taliban, slowly started to return to school. It was so strange to see them walking to school, wearing their black uniforms and little white headscarves. Because some boys used to curse at them and throw stones at them, the girls always traveled in groups for safety. My two sisters did go school for a short while, but not for long. My father deemed it too unsafe. I was happy about this. Not because I wanted to deny them an education, but because I didn't want any of my friends telling me they had seen my sisters outside.

There is a religious *hadith,* or saying, that discourages gossip. It states that gossiping is *haram,* forbidden—*as though you eat the flesh of your brother.* But the reality is that it is rife. And most gossip in our community was about females—whether they deserved it or not. So damaging could this be, that if anyone says anything bad about a girl, harasses her, or accuses her of misbehaving, some families either go and kill the person responsible for the slur, or move house or, in even more extreme cases, murder the girl, regardless of whether she was innocent of all wrongdoing. Either way, taking some form of action is seen as essential; otherwise the family reputation is destroyed.

As I didn't want any harm to befall my sisters, I genuinely thought that their staying away from school and safe from all of these risks was the best option. I was still as pious as I ever was; in some ways, even more so, since the arrival of the "foreigners."

EVERYTHING WAS STILL NEW, STILL UNSETTLED. THE WHOLE country was in a state of uncertainty. There were fights between different rival groups, some of whom supported the old Taliban regime, others who were now siding with the Americans.

The idea that the Taliban left and suddenly everything in my country improved overnight is nonsense. To many people, like my grandfather, it felt like the world had collapsed into immorality.

And as bad as the Taliban might have been, the NATO forces were far from perfect too. I witnessed one incident, which, like the stoning, continues to haunt me.

We were driving to Jalalabad, the nearest city to us. U.S. armored vehicles were up ahead and the road was blocked on both sides, meaning no one could pass. Suddenly, a car full of people, including children—swerved out and drove toward the checkpoint from the oncoming side. I don't know why they did this—maybe someone was sick, perhaps they had another kind of emergency—but for whatever reason, they drove on.

There was a hail of gunfire from the American soldiers.

The car burst into flames, killing everyone inside the vehicle. I remember a silent scream from the car filling my head as my father held me to him and tried to shield my eyes.

"Why? Why, why, *why*?"

There were so many incidents like this one. We used to hear the stories—accidental bombings of weddings, killings of innocent farmers, U.S. soldiers shooting anyone they thought might be Taliban. People disappeared into the hands of the Special Forces and were not seen again for years. Women waited for husbands who never returned. Their crime—who knew?

There were also the petty humiliations of life under occupation. There were near-constant roadblocks. One poor woman gave birth on the side of the road because the soldiers refused to let her and her husband past.

I hated the Americans, because they came to conquer us. I saw these latest invaders as worse than the Russians.

Politically, though, things became stable enough for Uncle Lala to return; it turned out he had been sheltering in Pakistan. He offered his services to the newly democratic Afghan government. The new provincial governor knew him well and knew what an intelligent man he was, and he gave him a very senior role running the prison service. But some local people feared and mistrusted him because he had been such a key figure in the Taliban. Others were jealous of my family's success, and saw denouncing my uncle as a way to get back at us. The Taliban were still thriving, though largely underground, in secret meetings and bunkers. The U.S. military was hungry for information, and people used old rivalries as an excuse to inform on others, whether or not the accusations were true. Because of my uncle's history, it became impossible for him to stay in Afghanistan, so a few months later he left again, again for Pakistan.

The U.S. forces had a large base near our house, and they knew Lala used to be head of Taliban intelligence for the East. They believed he had gone back to join the organization in a mountain hideout somewhere, and came regularly to our home to interview my father and uncles, even though my father didn't know where my uncle had gone—he hadn't heard from him since he'd left.

The men of the house, including me and my brothers, despised these visits. The soldiers had no understanding of our culture and stared directly at the women instead of turning their faces away. My little sisters Taja and Razia were so scared by them they cried.

"Gulwali, be polite and answer them," my father used to insist. "Better to not make trouble for ourselves."

But inside, I seethed.

"YOU ARE A YEAR OLDER NOW. BEHAVE LIKE IT." MY MOTHER'S voice betrayed no maternal kindness, but after she spoke she broke into a rare smile.

It was my eleventh birthday—October 2005. In the Afghan calendar it was 1385. The Afghan calendar is based on the date that Islam became predominant in Afghanistan, around fifty years behind the main Islamic calendar.

We didn't celebrate birthdays. People say it's unIslamic to do so, but really it's more of a cultural thing: many Afghan kids from poor or uneducated families don't know what day they are born; if your family is educated, they will usually write it down somewhere so you know. For my mother to make reference to my birthday was a special occasion. My chest swelled and I spent the rest of the day strutting around the compound.

During the occupation, our family continued to grow. My mother gave birth to another son, my younger brother Nazir. My parents were delighted, and his arrival softened their pain over losing their twin boys. But he was a sickly child who didn't eat much; I recall my father often injecting him with some kind of serum to help his appetite.

Around this same time, about three years into the occupation, my mother's parents, along with my grandfather's second wife and the fifteen children he had with both wives, returned from Pakistan, settling in a district about six hours' drive away. A deeply devout man, my mother's father was fiercely opposed to violence of any nature, and my mother thought it was a good idea for me and my brothers and sisters to stay with them, both as a precaution and as a way to get to know her side of the family. We were out of school, which felt like a treat, and it was fun spending time with this new set of grandparents and my other cousins.

Until one night news about my father and grandfather reached us.

We had just eaten a delicious dinner of home-reared lamb and rice. My mother's father took Hazrat and me outside. His voice was shaking, yet his words, in the traditional Afghan way, were brutal, unforgiving, painfully matter-of-fact: "Boys," he told us, "your father has been killed. You have to be strong. You are the men of the house now. Your father is gone."

I recall my legs buckling underneath me, but after that I can't remember anything. I think I have blocked it out—my mind can't go back there.

The details of my father's death are sketchy, based on the little I was told. My mother didn't want us to know the full horror of what had happened, and I didn't ask. I still don't want to learn the full truth, because knowing the details will never bring my family back.

But here's what I do know. The night of my father's death, U.S. troops again visited our home. There had been an attack on their base that had resulted in several soldiers being killed, and they suspected the weapons used in the attack had been stored in our house. They were angry, and this time they didn't ask questions: they came with snarling dogs, kicking in the doors. They searched everything—even the women—throwing furniture, personal items, and the Quran onto the floor.

They came ready for a fight. And they got one.

Neighbors, relatives, and extended family came to help drive the invaders out, all of them armed. By the time the shooting stopped, five of my relatives had been killed, including my father and my beloved grandfather. My two uncles survived, but they were arrested, thrown into prison.

I honestly don't know if the weapons came from our house or not. I was just a child, and these things were never talked about in front of me. Yes, my family were Taliban sympathizers, but I know my father would not have been happy putting his children at risk—he was never really happy when former Taliban members came to our house. But to refuse these visitors would have brought even more trouble on the family. The reality is that in war, ordinary families like ours are often left to make hard choices: appease one side and you make an enemy of the other side; try to placate the other side and the first side wants to know why. You can't win. The best you can do is try to negotiate a middle path and keep everybody happy, thus best ensuring those closest to you are kept safe. If indeed this was my father's strategy, and I will never know for sure, then tragically it failed.

I knew only one thing clearly. I had lost my father and grandfather, and so, as was the tradition in my culture, I wanted revenge.

CHAPTER 4

LOSING MY FATHER CHANGED EVERYTHING. OUR FAMILY was plunged into turmoil. My mother and my little siblings came to join us at her parents' house. My aunts also returned to their respective parents' houses.

It wasn't just the shock of losing my father—money was in very short supply, too. In countries like Afghanistan, the reality is that if the breadwinner is killed, the family he leaves behind will be very hard put to live.

One solution was available to us: my father's big ledgers from his doctor's surgery—the ones detailing the debts owed to him for medical treatment. It fell to my brothers and me to call at the homes of the people listed in them, and ask, demand, then beg for the money my family was owed. I detested it. Often these families were in worse straits than our own, yet the ledger had it all there in black and white: it was where our next meals would come from.

After a while, we went back to live in Hisarak, where we'd lived before the invasion. We had a huge extended family and people gave us a lot of support, with different relatives—including the husbands of my dad's sisters—coming to stay and support us. But with the men of our immediate family dead, imprisoned, or, in the case of Uncle Lala, missing, it was left to my brothers and me to become the men of the house.

Other problems began to crop up, too. Taliban representatives began to visit more and more often. They wanted my brother and me to become fighters or even suicide bombers—martyrs—to avenge our father's death.

I was so angry that I wanted to do it. Hazrat and Noor wanted the same. The three of us didn't really talk about it but I think we had an innate brotherly understanding of each other's pain. But my mother knew that an angry and hurt child couldn't understand the consequences of such a drastic action and could be easily manipulated in his grief. She had been there for the horrific event and she too was filled with fury and pain—certainly a lot of Pashtun women would have wanted their sons to take revenge, even if it meant them losing their lives. Revenge is a central yet often lethal part of *Pashtunwali*—if you don't avenge yourself against your enemies, according to tradition, you have failed as a man.

But my mother was influenced by a different set of thinking—her deep and abiding faith. Thanks to my maternal grandfather, she had a genuine and strong understanding of true Islamic law, a religion that prohibits the taking of life. She had been taught that killing anyone was wrong, even if the reasons might seem justified. She had been taught to try to forgive, to show compassion, and to accept tragic events as God's will. It wasn't easy for her, but her faith helped her get through losing my father.

My mother was also scared for me and Hazrat, because the U.S. forces had also turned their attention to us, urging us to become informants. They wanted information about my uncle's contacts and where Taliban weapons might be stored. She feared that if we followed that route and got involved with the NATO forces, or even the Afghan authorities who were cooperating with them, then we would be seen as traitors and killed.

It was a genuine fear. We knew other families where this had happened.

The approach from the U.S. forces was all very carefully done: there wasn't really any direct contact. It was more indirect—through letters or messengers. We had a family friend who revealed himself to be a U.S. informant. He had survived so far because he was a very powerful man—so powerful, we'd been told he had the ability to call in U.S. airstrikes if he needed to.

Hazrat and I visited him at his house to try to find out information about my uncles: we still didn't know to which prison they had been taken or what their charges were.

We sat on an ornate gilded French-style rococo couch and drank tea. "Do you know the saying, boys? 'The enemy of your enemy is your friend?'" Of course we did. It was a very common Pashtu saying. "You would be wise to remember it now. I will do whatever I can within my power to support you and help you, but you would also be wise to think of who your new friends might be."

I was suspicious of him. As we all knew he was working with the U.S. military, advising them on local intelligence issues, I was worried that he could be double-crossing us. But I also knew he'd known our family for a long time and had known my father very well, so I hoped that stood for something.

At the end of the meeting, he warned us both to be careful, telling us flatly that the NATO forces wanted us to work with them. He said if we didn't do what they said, there would be a danger we'd be killed or end up in Bagram prison. They had impunity. He knew it and we knew it. He also made clear that if we made any moves toward the Taliban side, he would know about it.

And, in that circumstance, he was more than clear that he couldn't and wouldn't help us.

IN THE WEEKS FOLLOWING OUR FATHER'S DEATH, OUR HOUSE was full of visitors, all with lots of advice, and all of it conflicting.

Senior elders told us to get involved with the Taliban, insisting it would be for our own safety. A messenger arrived, a stranger, who said he was there to offer condolences. He said that he was representing the Taliban shadow governor for the province, a man who had been appointed by Mullah Omar himself. He told us the Taliban authorities were happy to take revenge on our family's behalf, but that before doing so he needed my brothers and me to fight alongside the Taliban.

"Work with us, not against us. Join us to fight the infidels and expel them from Afghanistan."

He went on to say that he understood we had family duties and could understand why we might want to stay out of things, but he still hoped we'd play our part.

"This isn't just about revenge for your father, boys. We want you to be part of a bigger and greater mission—expelling the invaders."

This message made sense to me. Of all the different people advising us, he was the most persuasive—to me, at least. Hazrat never liked him.

Later, the messages took a more sinister and threatening turn: "Night letters"—handwritten notes tossed over our compound wall or through the doorway. The handwriting was always different, but the words were always the same: *Be martyred, or die.*

On our way back from school, bearded men on motorbikes followed us home. Some of them were cajoling, others directly threatening; some of them even tried to grab us and make us go with them to a Taliban training camp there and then. It got to the point that as soon as we heard the sound of an engine we ran to hide.

So many different people offered protection and support, but we didn't know who we could trust.

I doubt that the Taliban would have killed us for refusing their invitation; they were simply trying to scare us into joining their cause. But I do think if we had collaborated with the U.S. military, they certainly would have killed us for that. My brother and I were just two little boys, but in the political and moral whirlwind of that time, we were pawns in a deadly game of chess: easily played and easily lost.

In the end, my mother decided our fate. One night, she sat Hazrat and me down. She was very composed. "Gulwali, Hazrat, you need to leave. You have to go somewhere far from here, where no one knows you. Noor is too little, so he will stay with me. We are working on a plan."

That was it. She didn't say where, for how long, or when. She didn't ask our opinion or how Hazrat or I felt about it. I wanted to ask a hundred questions but she was already busy making tea, her tight-lipped look making it clear the conversation was over.

Never did I imagine that her plan would involve paying human smugglers thousands of dollars to take us to Europe.

IN MY HEAD I THOUGHT THE JOURNEY—AS MUCH AS I KNEW about it—would only take a few weeks.

Hazrat and I were sent first to my maternal aunt's house in Waziristan. I assumed we'd be coming home at some point, so there were no big good-byes, no tears of sadness. Even when I hugged Grandma and my little siblings good-bye, I didn't make a fuss, because I thought I'd see them all again soon. We'd been moving around so much since the conflict began, none of it seemed unusual. And I didn't question my mother, because in my culture, as a child, there are things you just do when you are told to—you don't ask why or require an explanation. We never thought about ourselves individually, just about the family.

My uncle Qais (the husband of my mother's sister), came to collect Hazrat and me. The journey to Waziristan took us across the lawless heartland of the Pashtu tribes, a place that has been described as the most dangerous and controversial border in the world. Although recognized internationally as the western border of Pakistan, the Afghan government does not recognize it. It's known as the Durand Line, named after Sir Mortimer Durand, a British diplomat who, in 1893, after two Anglo–Afghan wars, negotiated the 2,640-kilometer-long boundary between what was then British India and Afghanistan. The idea was to create a neutral buffer zone to improve diplomatic relations with Afghanistan and limit Russian expansion—a battle for the control of central Asia known as the Great Game. Pashtuns refer to the Durand Line as: "a line through our heart." For us it is a remnant of colonial oppression, and the sense of injustice had been instilled in me for as long as I could remember. Pashtuns see the real border as further into what is now Pakistan, where the river Indus separates both the Pashtun and Baloch lands—Balochistan—from Punjab and Sindh. The Baloch people, who the Pashtuns see as cousins, have been fighting for independence from Pakistan for nearly sixty years.

As we journeyed over mountains, rivers, and lakes, my heart stirred, both with the pride of being a Pashtun, and with the happy, nomadic memories of my early childhood. How I longed to be in the fields again, tending sheep with my grandfather. I was still grieving so very badly for both him and my father, the two men I had loved so much.

IT TOOK OVER A DAY AND A HALF OF TRAVELING, AND I WAS very happy to reach Waziristan. It was so exciting. The town was ringed with beautiful, snow-covered peaks crisscrossed with sparkling rivers.

Uncle Qais took us straight to the main bazaar. It was filled with merchants, all of them shouting their wares in a guttural tribal dialect I could barely understand. They crouched on the ground drinking green tea, arguing loudly, or comparing wild birds kept in long rows of cages. Most of the shops were either selling guns, making guns, or testing guns—by firing them into the air. Every time a gunshot rang out, I jumped. Everywhere there were semi-wild street dogs, while the smell of sizzling kebabs, sold by roadside vendors, made my stomach rumble with hunger.

I was mesmerized, particularly by the piercing green eyes of the locals. The people of that district are famous for their beautifully colored eyes. Even when they are happy and at peace, they manage to look fierce and otherworldly.

My aunt's house was small and surrounded by fields of opium, wheat, and cannabis—the three main crops of Waziristan. I was desperate to run free in the fields, but it wasn't safe because the Pakistan military were fighting local Taliban militants nearby. Mostly my brother and I sat in my aunt's house, doing nothing, getting bored, and feeling homesick, annoyed at not knowing how long we would be there or what was going to happen to us. We were only really allowed out to go to the village mosque to pray. I loved being there, as it helped settle my mind. I prayed that all of the pain to hit my family would serve a purpose: I was beginning to understand that life was a test, and that it was important not to give up.

After a week or so, my uncle took us to buy clothes. I was furious when he made us try on pairs of jeans. "Uncle, I'm not wearing these. It's what the invaders wear." At home, we only wore the traditional *shalwar kameez*. I was so anti-Western, it felt like a complete betrayal to even consider wearing these devil jeans. I couldn't understand why he was making us do it: denim was the cultural uniform of my enemy.

But my uncle was in no mood to argue. "Put them on. Do you think I am wasting my money here?"

Sulking, I did as I was told but I hated the way the stiff cotton felt on my skin. He also bought us a small rucksack each, T-shirts, underwear, woolen gloves, strong boots, and warm jackets. By the time we'd finished shopping, the rucksack was full.

No one had yet spelled it out to us, but it was fast becoming obvious what might be happening.

Hazrat and I were far closer now than we had been in our early childhood. The trauma of the past couple of years had bonded us, but like all brothers we still got on each other's nerves. Things weren't helped by the fact that even though he was a year older than me and very clever, I always saw myself as the leader.

"They're sending us away, aren't they?" I said to him that evening.

He nodded.

I saw how he was trying to look wise and mature about it, but I blurted out the only feeling I had on the subject: "Let's not go. Let's go home. We're needed. We run the businesses and look after the women. I can take over the tailor's shop. I'm a good tailor."

"*No,* Gulwali." My brother looked at me. "This is our mother's decision, and we must respect it."

"She's wrong," I insisted.

"Listen, if she's made this decision, it has to be serious."

"You're a *khazanouka*." Calling my brother a feminist didn't help matters, and I still felt angry. It wasn't just my mother I was upset with; it felt like the whole family was involved in this plot.

That feeling only intensified when my uncle took us to a kebab restaurant the next day. He greeted the owner of the restaurant, who took us to a back table away from the crowds.

"So, these are the boys?"

"Yes. And they will travel together? Their mother insists on this."

"You have my word," said the man.

I sat there nervously listening to my uncle and the man debate money, paperwork, processes. My uncle and the man talked about us as if we weren't there. Maybe it was too dangerous to share information, or perhaps they didn't want to scare us even more than we were. I don't know. I just remember feeling nothing but confusion and despair.

IN LATE OCTOBER 2006, WHEN WE'D BEEN STAYING WITH MY auntie for around three weeks, my mother arrived for a visit.

I was happy to see her as I'd really missed her, but at that time I was also very angry with my mother. I felt betrayed. I had been the child who had prided himself on responsibility and protecting our family honor: it was a role I relished. Being stuck in Waziristan, I felt like I no longer had that status. And, unfair as it was, I blamed her.

She kissed me on the forehead, then went straight to talk to my uncle. The pair of them sat talking in hushed tones, and I wanted to shout at them. Why were they doing this?

The next night, Hazrat and I were told that we—my mother, aunt, and uncle—were going to visit someone's house for dinner. This was to be expected: if you are visiting a new place, local people will invite you for dinner—it's a key part of *Pashtunwali*—so I didn't think anything out of the ordinary.

The house was about a ten- to fifteen-minute walk away. My mother and aunt walked alongside us, both wearing their blue burqas. On arrival, Hazrat and I were ushered into the main quarters, where a few men of different ages sat on striped kilim-covered cushions on the floor, drinking green tea.

As I stuffed some naan into my mouth, I listened to the men debate politics.

"The war over the border has reached us here now."

"The government will never win this."

"*Pashtunwali* itself is at stake."

"These foreign fighters need to go home. It's because of them the Americans are dropping drones on our head."

"How can you say that? The Americans are the ones who occupied our brothers' country."

"Yes, but isn't it because of them that these Chechens and Arabs are in our midst?"

"These Chechens and Arabs are our guests. They came here to fight the holy war. They are our brothers."

"Some brothers. Marrying our women and bringing danger to our villages. What good is this doing for our people?"

"There are thousands living in camps now. We are refugees in our own homeland."

After the food and the debate were over, we drank hot, sweet green tea and ate dried apricots, mulberries, and almonds. I was beginning to feel relaxed and sleepy when my uncle and the host stood up and told Hazrat and me to follow them.

They walked us to the door, where my mother was waiting for us. I was aghast to see she had our rucksacks next to her. I had no idea who had brought them here. It didn't feel right.

Hazrat looked at me, as shocked as I was. I glanced at my mother for reassurance.

As she stood there, her knees seemed to give way slightly. I thought she might fall down.

She looked at my brother and me. "You must hold on to each other's hands. Never let go of each other. Do you understand?"

"What? *Morya.* I'm not ready. Please—"

"I said, DO YOU UNDERSTAND?" She stared at us so hard through the mesh of her burqa I thought her eyes would burn through my skull.

We both nodded.

"Be brave. This is for your own good." And then she said something that froze my heart. "However bad it gets, *don't come back.*"

I STARTED SHOUTING. "I AM NOT GOING ANYWHERE. I AM *NOT* leaving my family. I am needed here."

"Gulwali, this is for the good of the family. Do as we say."

"No."

I looked over at Hazrat. He looked stricken.

My mother started speaking again. "Listen to me, both of you. These people will take care of you, but you must take care of each other. Hold each other's hands. *Always.*" She looked at us both again, hard, holding our eyes. Then she looked down, taking something from the folds of her burqa.

She handed us each US$200; my uncle gave us a handful of Iranian currency.

Our host looked at his watch. "We should move. Long drive ahead." He began to herd us outside, to where a car was parked.

I started to feel sick. "*Morya,* I don't want to go. Let's go home now. Please take me home."

She glared at me and, for a split second, I thought she was going to hit me.

"*Morya,* please." I said, absolutely desperate now.

The man shifted impatiently, looking at his watch again. "We need to hurry, madam."

My mother began to shake. I could see her eyes through the mesh of the burqa flitting from the man and back to us.

"Where are we going, *Morya*?" I asked.

"To safety. To Europe."

"They will be there in a few weeks," the man assured my mother. "It will be like a holiday for them, an adventure." And with that he literally bundled us into the car.

My mother stood watching as we got in, standing tall, proud, and resolute. I don't know how she would have reacted as the taillights of the car rounded the corner: Did she collapse, screaming and howling, finally allowing the trauma and grief of losing her husband and now her two older sons to finally flood through her?

More likely, given what a strong woman she was, she simply stayed quiet and locked her pain deep inside.

I WAS TOO TRAUMATIZED TO SPEAK. MY ONLY COMFORT WAS my brother's hand gently squeezing mine when he noticed my tears threaten to fall.

"We'll be okay. We're together. I am going to look after you."

That night we were driven to a city. I think it was Peshawar, but no one told us. We were taken to stay with a family, complete strangers, but given food and a pair of mattresses to sleep on.

In the morning, two new people arrived: a man and a woman. They were Dari-speaking Afghans.

Dari and Pashtu are the two main languages of Afghanistan. Pashtu is my mother tongue, and I'd lived in an exclusively Pashtu area, so my Dari was in no way good, but I could make myself understood—most Afghan kids speak both.

The woman explained that they would take us to the airport and fly with us to Iran. She told us that, in the airport, we were to be on our best behavior and pretend they were our relatives. She told us to make no noise and not to attract the security guards' attention. From there we would go to Europe.

"Where in Europe?" I demanded to know.

The woman shrugged. "I don't know. Don't ask me. I'm just an agent. Your family made the arrangements with people in Kabul." She reached into a bag. "Here. Take this."

She handed my brother and me an Afghan passport each. I had never had one before. I stared at it in surprise as my photograph peered back at me from the page.

When had my family arranged this?

On arrival at Peshawar's international airport, I momentarily forgot my sadness. It felt so huge, with people, cars, queues, and machinery everywhere, and I stood gaping in wonder at the scanning machine as our little rucksacks went through it.

I was brought back to reality with a jolt when the man we were with picked up Hazrat's bag and led him through the gate. I went to follow, but the woman held me back. Hazrat didn't give me a backward glace—I assume because he thought I was right behind him.

I wanted to yell his name but the woman's earlier warnings about not attracting attention still loomed large in my mind. "I want to go with him." I spoke to her as forcefully and in the best Dari that I could muster. I had to make her understand what I said.

She ignored me, grabbing my arm so tight it hurt. I started to cry and she looked around nervously. "Be quiet." A guard glanced over in our direction. "It's okay, it's okay," she said, quickly. "Later. You will see him on the plane."

I stopped crying because I believed her. But when we boarded the plane, my brother and the man were nowhere to be seen. "Where is he? Where is Hazrat?" I demanded.

"Quiet," she hissed. "Different plane. You will see him on the other side."

This time I knew she was lying.

BY THE TIME WE ARRIVED AT OUR DESTINATION, I WAS A TERrified, lonely, and sobbing mess. As we went through Immigration, the reality that I was now in a foreign country hit me. Waziristan and Peshawar, although technically in Pakistan, still felt familiar because they were Pashtun areas. But Iran was completely alien.

Outside the airport, the woman simply pulled me over to a taxi without saying a word. I was begging for my brother and desperately straining my eyes in all directions in the hope that I would see him.

The female agent and the taxi driver spoke in Farsi, the main language of Iran. It's very similar to Dari—the two languages are virtually the same, though the dialect and accent are quite different. But to a Pashtu speaker like me, Dari was hard enough, and my mental anguish and their strange accents made it hard for me to understand anything.

Trying to get her attention, I tugged at the woman's sleeve, but she flicked my hand away as if I were an annoying fly. Still weeping, I was bundled into the back of a taxi and driven out of the airport into the surrounding countryside. The woman walked off without so much as a backward glance. I wanted so badly to see my mother or my grandmother—to have one of them hold me and reassure me and tell me it had all been a bad dream. I prayed that Hazrat was going to be wherever it was they were taking me.

The taxi driver, a skinny man with a big, hairy nose, glanced over at me as I sat on the backseat, snot and tears dribbling down my dirty, exhausted face.

"Welcome to the holy city of Mashhad."

This made me furious. Iranians were Shias. Muslims are split into two main branches—the Sunnis and Shias. My family was Sunni, and I had always been taught in the local mosque and at school that the Shia weren't true Muslims, something that as a child I had accepted unquestioningly.

But I had to admit that Mashhad was very beautiful. It looked new and shiny and clean; the pavements were so sparkly they seemed to glow. And the tall skyscrapers and shopping malls I had seen in books were everywhere; even the roundabouts had decorations in the center of them. I marveled at one that had a pair of hands holding the Quran carved in pink marble.

As I peered out of the car window, I saw an Iranian family walking down the street. The parents were holding the hands of their two children, a little boy and girl. The girl was wearing a white, lacy dress with a stiff petticoat and frills, while ankle socks and black patent shoes adorned her feet. The little boy had on a smart black suit and a red bow tie. The parents were clasping their kids' hands tightly and swinging them up into the air. The sight of their carefree happiness cut like a knife as images of my little sisters and brothers rushed into my head. Quickly, I blocked them out. I was already learning that in order to keep going I had to stop loving, stop remembering. Thinking about those I had left behind was too painful.

I was taken to a smart little hotel in the center of the city. The manager came out to the taxi and asked if I had luggage—I had to concentrate really hard to understand the question. I answered by shaking my head and clinging on to my rucksack: it was all I possessed in the world.

Once inside, the manager took me to one side. "Are you the person of Qubat?"

I had no idea what he was talking about. What was "Qubat"?

"Where is the other one? I thought there were supposed to be two of you?"

"My brother? Is he here?"

"Maybe. There are some others belonging to Qubat here. He might be with them. Let me show you the room."

He led me along a hallway and into an elevator. As he pushed some buttons, it began to move. I was fascinated—I'd never been in one before.

"Okay, you guys are in room fifteen," the manager said, jangling his keys as the elevator door opened.

Outside the room, he knocked on the door. Anticipation ran through me. I prayed Hazrat would open it. Instead, a skinny, tall youth with protruding teeth and messy hair poked his head round the half-opened door. "Yes?"

"The new arrival is here. He's little. Introduce yourselves and look after him. He's asking for his brother."

The youth gave me a toothy smile and introduced himself as Mehran. "*Salaam alaikum,*" he said, welcoming me in Pashtu.

I smiled back with relief. From that time on, being around other Pashtuns would always make the experience of being in a strange country feel a little bit easier.

The room had four single beds in each corner; in the center was an Iranian-style carpet with a peacock design. Two other figures sitting on the floor got up as I walked in. They both came over and hugged me. One, a clean-shaven, fit-looking man, who looked to be in his late thirties, introduced himself as Baryalai. The other said his name was Abdul. He was in his mid-twenties and seemed a little bit shy.

"You are too young to be here, little man," Abdul said. "Let me get you some tea."

"No, no, I'll get it," said Mehran.

This made me smile—typical Afghans, fighting over each other to show hospitality. These three made me feel instantly welcome.

As we sat on the rug drinking tea from tiny Iranian-style cups and saucers, I learned more about them. Baryalai was from Nangarhar, the same province as me, though he had grown up in a refugee camp in Peshawar. Mehran was also from Nangarhar and from Hisarak, the same district I'd grown up in. Abdul was the son of an army officer from Kabul.

They explained they'd traveled by coach from Kabul to the city of Herat in western Afghanistan. From there, they had crossed the border into Iran, where they swapped coaches to travel here.

We didn't really get into the details of why we'd all had to flee; I think there was an unspoken assumption there must have been good reasons, otherwise why would we be there?

"Have you eaten yet?" Baryalai smiled at me warmly. There was something about him that made me feel safe.

I hadn't eaten since Peshawar. I was starving.

"We have some food. Come, eat." Mehran was already piling fried rice from a steel container onto a plate. The food looked reassuringly similar to Afghan food.

I had so many questions in my head: How long were we staying there? What was happening next? Just one passed my lips: "Where is my brother?"

"Your brother?"

"Hazrat. He should be here. Is he in a different room?"

Baryalai looked at me and frowned.

"There are ten other Afghans. They are with different agents—we are the only ones belonging to Qubat. But none of them are called Hazrat."

"But they promised."

"'They' make a lot of promises, Gulwali."

I went silent. I think they could see I was scared and depressed, so they suggested we go for a walk. I was surprised we were allowed to go out by ourselves, and I jumped at the chance.

We wandered through a nearby park eating watermelon. Finally, I was able to ask all the questions I needed, including one which had been perplexing me ever since I arrived at the hotel: "What is 'Qubat'? Is it Farsi for 'kebab'?"

They all roared with laughter.

"What's so funny?"

Mehran wiped tears from his eyes. "I thought I was stupid, but this kid is *really* dumb."

Baryalai spoke next. "Gulwali, 'Qubat' is the main agent from Kabul who we paid to get us here."

"But I didn't pay anyone."

"No, but your family did. Qubat is a people smuggler. A big guy. I think I met one of your uncles when I was there in Kabul. I heard him arranging for Qubat to take his two nephews."

"Is Qubat here? He'll know where Hazrat is."

"No, Gulwali." Baryalai's tone gentled. "As I explained, he's the big guy. He lives in Kabul. I only met him once. He has guys working for him over here. He organized it all, and he's got our money. He's got every rupee I ever had. That's all I know."

"But what about Hazrat?"

Facing me, Baryalai put his hand on my shoulder and stooped slightly so that he was looking straight into my eyes. "Little man, I think they lied to you. You have to forget your brother for the time being. We're your brothers now."

I didn't want to embarrass myself by crying in front of my new friends, so I bit hard on my lip.

But Baryalai had noticed. "Try not to cry. Your uncle paid for you both, so he'll be somewhere on the same journey. I am sure you'll meet him again soon."

CHAPTER 5

EARLY THE NEXT EVENING, WE LEFT THE TEMPORARY SEcurity of the hotel. Baryalai told me we had a long coach journey ahead, driving from Mashhad to the city of Esfahan, then on to the Iranian capital, Tehran.

While I was reassured by my new traveling companions, I still felt very lost and scared. Every time I thought of home, my grandmother, or my little brothers and sisters, I couldn't stop my tears. And when I thought of Hazrat—taken God only knew where—I felt physically sick. My mother's final words to us, "Hold each other's hands always," rang through my head. She would be beside herself to know we'd been separated.

The hotel manager drove us to the coach station and bought us our tickets. The bus tickets had assigned seats; ours were located in the long back row.

Baryalai wasn't happy. "This is dodgy. If the police come on board they will see us straight away. Get us a different seat."

We'd entered Iran legally on tourist visas, but if we or our passports arose suspicion at any point from now we'd be rumbled for the refugees that we were. If the police asked us what sites we wanted to visit or where we were going, how could we answer? We'd have been arrested for sure. Iranian prisons are full of Afghan refugees. This was common knowledge.

The manager looked exasperated. "It was full. These were the last seats."

"So we'll take the next bus," insisted Baryalai.

"It's not for six hours. They are expecting you in Tehran tomorrow evening. If you delay, you'll miss your pickup."

Baryalai still wasn't pleased.

I didn't want him to be sad, so I tried to reassure him: "It's okay. We'll manage."

But as the coach left Mashhad for Tehran, I felt as if I was headed in the wrong direction, further and further away from everything I knew.

I WAS SMALL ENOUGH TO FIND SOME SMALL MEASURE OF comfort on the vinyl seats of the bus. When not sleeping, I snacked from a packet of salted sunflower seeds and watched as Iran unfolded outside the window. It looked so clean and organized. Unlike back home, the roads were smooth and asphalted, and I marveled at all the modern-looking buildings festooned with strings of colored lights.

Every few hours the driver stopped to refuel, or for meals. We would pile out, grateful to stretch our legs and visit the bathroom. At some point during the night, I woke as the bus pulled into a larger terminus. Dozy and confused, I stumbled into the cool air. The depot was overflowing with tired travelers blinking away sleep. Their clothing and hats marked them out as decidedly Persian. There were few women around, but those who were tended to fall into two categories: the older ones wore full black robes that covered their heads and bodies but left their faces showing; the younger ones wore jeans and long tunics or belted coats, with brightly colored scarves covering their hair. I found it unsettling to see so many women outside without their faces covered. I was still thinking of my province, where women were rarely seen outside the home, and when they were, they wore the traditional blue burqa.

The depot was full of enterprising stall owners, all selling their wares to hungry and thirsty travelers. Decorated stalls sold rows of tempting dates, mangoes, apricots, and oranges; others displayed myriad different-colored marzipan sweets and decorated little cakes, along with cans of Coke and cups of sweet black tea. Walking between the rows, looking at the mouthwatering wares, I became entranced. I quickly lost track of time. I came to with a sudden realization: the driver had told us to be no more than ten minutes. I ran back to where the coach was and opened the door with a puff of relief. Only to see an unfamiliar driver. Panicking, I stepped back and looked left and right. There was row upon row of identical-looking coaches. Where was mine?

My heart was pounding. My passport was on the bus, along with most of my money. I ran up and down the rows of coaches searching for my bus, but they all looked the same.

My face was hot and tearful; I pushed people out of the way.

"Hey, kid, watch what you're doing."

I glared back though teary eyes. "I've got to find my bus," I said, forcing my way through a gap in the crowd. Blind panic was setting in. What if they'd gone without me? Surely the others wouldn't let that happen.

My tears were flowing so much now I could hardly see; my heart was pounding so hard I thought my chest would explode. Where was I? I didn't even know the name of the town we were in.

I had no idea how to survive in a foreign country. I was just a little boy. How could this be happening to me? Where would I go? What would I do?

I let out an angry sob. "*Morya*. I want my *morya*." In my panic, I continued to run, shoving my way past the back of the next bus until I smacked right into a familiar form.

Mehran looked truly furious. "Gulwali. Where the hell have you been?"

"I got lost. I'm sorry. I thought you'd gone."

"We nearly did. The driver wanted to leave. Come on, we have to get back on now."

As I slumped into my seat, I could barely look the others in the eye.

"Are you all right?" asked Baryalai.

I couldn't show them weakness. "Yes. Fine."

"Make sure you stick close to us in future."

I tried to rest as the bus roared through the night, and eventually sank into a fitful sleep.

When I next opened my eyes, I thought my journey to freedom might be over before it had begun: an Iranian soldier was sitting right next to me. I opened my mouth to speak, but Baryalai quickly spoke for me. He was talking in Dari, the Afghan language very close to the soldier's Farsi. It meant that it was obvious we were Afghans, but at least he and the soldier could converse easily.

"Good morning, Gulwali." His eyes implored me not to say anything.

I couldn't quite work out what was going on. The solider was chatting to my friends casually, as if it was the most normal thing in the world. If he'd come to arrest us, why hadn't he already done it?

I sat as quiet as a mouse, not daring to say a word.

As they conversed, however, I realized the soldier wasn't there to harm us. It turned out that he'd just finished his duties and was on his way home to see his parents, just a bus passenger like us. He asked us polite questions about how we were finding Iran and did we like the scenery. He seemed very proud of his country. I guess he just assumed we were on holiday, or working there. It's not unusual to see Afghans in Iran; many Afghans go there for business or to find work in the Iranian construction boom.

As nice as he was, though, I was still very relieved when he got off the bus.

Later on in the morning, we stopped again for breakfast, pulling into a service station. There were a couple of assorted restaurants and a small mosque. This time I made sure to stay with the others.

At the ablutions area, a row of benches and taps was surrounded by men preparing themselves for prayer. Ablution is the ritual Muslims do before prayer, which involves washing the hands, arms, face, and feet. We joined them as the *azzan*—the Islamic call to prayer—stirred my soul. Even as a small child, it was the most wonderful sound in the world, filling me with a sense of peace.

I had been taught from a young age to fear the Shia as apostates—yet the *azzan,* the ablution . . . everything seemed to be the same as my own rituals. As we went inside the mosque to pray, I realized that we also said the same prayers. The only real difference I noticed was that when I prayed, I placed my hands in front of me; they had their hands by their sides. They also prayed out loud, whereas I was used to worshiping silently. But, apart from that, everything seemed normal.

But there, in the mosque, as I looked around me, I did not see heretics or disbelievers, as I had been led to believe Shias were. All I saw were men of faith. And I was thankful to be there, praying by their side.

WE ARRIVED IN TEHRAN EARLY THAT EVENING, AFTER A FULL night and day on the road. Compared to cities like Peshawar and Kabul, it was a hugely modern and clean place. There were endless concrete suburbs. The traffic was bumper to bumper, with yellow taxis everywhere. Diesel fumes made the air chokingly thick.

The moment we stepped off the coach was chaos. The terminus was massive, and as we descended the bus steps, taxi drivers were grabbing at our sleeves.

"Taxi."

"Come with me."

"This way, sirs."

It was so bewildering. We'd been told someone would be there to meet us but we had no idea what they looked like or who they were. Fortunately, we did have a phone number, scribbled on a piece of paper the hotel manager had given Baryalai.

We managed to break away from the taxi touts and find a public pay phone.

"I'm here. Outside the station. Come out." The man's voice rang out in clear Farsi.

We walked out into the street but still had no way of recognizing the man. It was up to him to work out who we were. It clearly took him a while because we were standing there nervously for twenty minutes before he found us.

A young man in jeans and a sports jacket approached us. "*Salaam*. Are you Qubat's guys?"

We nodded.

He apologized for being late, explaining there had been delays on the metro. I had no idea what a metro was.

"Let's walk along the street," he said. "The taxis here are too expensive."

As we followed him, I looked around and marveled at the city. The roads were lined with beautiful flower beds, all different colors. There was a lot of noise and traffic still, but the city seemed so calm and ordered. Most consumer goods in Afghanistan come from Iran so I had had an idea it was a developed place, but the reality was something else. Our capital city of Kabul was once a famous tourist destination known for its gardens and parks, but the years of conflict had damaged it terribly. Although still very beautiful in its own way, these days open sewers run down the sides of potholed roads, and rubbish piles up in the rivers. To me, Tehran was the epitome of what a capital city should look like.

We then got into a taxi, which took us through winding backstreets to our destination, a small traveler's guesthouse known as a *musafir khanna*. It was located above the top of a hardware store in a busy bazaar. Inside the *musafir khanna* were already a few Afghans. And it was clear they weren't happy with the guesthouse owner, even though he was also Afghan.

"We were supposed to leave Tehran two days ago, but this guy is forcing us to stay here," complained a large, chubby man who introduced himself as Shah. He was from central Afghanistan, home of the Hazara people. The Hazara make up about 10 per cent of the Afghan population, but throughout history they have been treated as second-class citizens and had their lands taken away, in part because they are Shia. Back then, all I really knew about the Hazara was that they had opposed the Taliban ideology, something that at the tender age of twelve I couldn't understand anyone wanting to do. Shah, I soon learned, was sharing a room with Faizal, a younger man from Balochistan in Pakistan.

As we ate dinner, Shah and Faizal told us more of their stories. Shah couldn't believe we four had all entered Iran legally. He'd walked and crossed the border illegally. "I didn't think I'd live to see Tehran. We left Afghanistan at Nimroz and trekked for so long into a city called Zahedan. The damned Iranians shot at us. I saw a child fall and die right in front of me. His brother was screaming. We were running like scared chickens."

Hearing this made me think my experience on the plane hadn't been so bad after all.

Faizal—a journalist—had also faced great danger crossing on foot from Pakistan into Iran. He was clearly very disturbed. "You don't understand what's happening to my people. The military are disappearing us. I found my cousin's body with wire round his neck. He was only nineteen. I left my whole family, my two kids, behind. I don't know when I will see them again."

The situation at the *musafir khanna* was really unsettling: the owner told us he was Qubat's business partner in Iran, but then started yelling at us. "Qubat was supposed to send enough to get you to Turkey. But where is the money? Where is it?"

None of us had any idea what he was talking about.

"I'm not buying your tickets out of my own pocket. You don't go anywhere till I get paid. It's bad enough you all have hungry mouths to feed." The manager pointed at Shah. "This fat Hazara has cost me too much money already. I should throw him on the street."

Shah threw his hands in the air and rolled his eyes, before saying an old Hazara expression: "Hazara also have God."

I liked Shah—even when he was angry he was still funny.

That night I lay on a lumpy, metal-framed bed watching cockroaches scuttle across the filthy walls. This place, which stank of boiled meat and stale cigarette smoke, couldn't have been more different from the smart, clean hotel in Mashhad. For a long time I stared at the ceiling waiting for sleep to come.

Even though the people around me were being nice, they were still strangers. I ached for my family. All I wanted to do was get back on the coach and find a way home.

But my mother's words, "*Don't come back,*" echoed through my head.

CHAPTER 6

"I. AM. NOT. GIVING. YOU. MY. PASSPORT."

Mehran looked so angry I thought he was going to hit the guy.

"None of us are handing them over," Baryalai chimed in.

Qubat's guy, the guesthouse manager, had come into our room to tell us his money had finally arrived and that we were to take a bus on to the city of Tabriz that night. But he'd demanded that we hand him our passports.

"If the police catch you with your passports, then it will go badly for all of us. You are going to the Turkish border and they'll know you want to cross. Better to turn them over to me, then there's no evidence for the police."

We didn't buy this argument. Any ID was better than no ID, fake or not. Surely we couldn't move around Iran without one?

"Maybe you want to sell them and leave us here to rot?" Mehran was becoming hysterical, yelling at the top of his voice.

"Shut up, you foolish boy. Do you want people in the street to hear us? It's for your own good."

"*No.*"

But the guesthouse manager was getting irritated by our refusal now. "You were supposed to be here for a few hours, but you stay for days—and now you try to tell me how to do my job. Don't you think I have done this hundreds of times before?"

Baryalai wasn't buying it. "I bet you've tricked men like us a hundred times before."

"A thousand times," muttered Shah.

"Fine." The manager finally snapped. "Don't give them to me. Stay here in Tehran and rot. No passport, no ticket. No Europe. Your choice."

This threat scared me more than anything, and I found the courage to speak. "What guarantees do we have that we'll be safe?"

But Faizal shouted me down. "Shut up, boy—don't speak of things you know nothing about."

"Leave Gulwali alone." Baryalai clearly didn't like Faizal.

"The boy is blind to everything except his little crush," retorted Faizal.

I glared at him in fury.

The manager saw his opportunity to reason with us. "The boy can see enough to know I am right. The only guarantee I can make you is that you don't go anywhere until you give me those documents. But you know what? Perhaps I will turn you over to the police and tell them you are refugees—you have cost me too much money already."

I couldn't help myself. "No, please, don't. We'll think about it, okay?"

"Think about it? You think you have a choice?" The man snorted with derision. "You really are wasting my time now. I am not an unkind man, but this is my home, and my rules. You are going to need fake papers—but that won't happen until you get closer to Turkey. Until then, you are safer not having any papers at all."

Nothing he said made sense. If we were caught without any documents, then surely we'd definitely be heading to prison. If questioned, I'd have no way of proving I'd entered the country legally on the tourist visa.

Mehran shook his head. "Look, these are *our* passports."

"'Our' passports? Weren't they provided by Qubat? These are his property, not yours. They served their purpose to get you here, but you don't need them now. Give them to me."

"No."

We were clearly getting nowhere. I didn't want to hand over my passport either, but nor did I want to stay in that stinking guesthouse a moment longer. "I don't think he's lying. If it's for our safety, we should."

A COUPLE OF HOURS LATER AND FOUR BRAND-NEW AFGHAN passports were in an envelope on the reception desk, weighted down by an illegal bottle of Johnnie Walker Black Label. We were absolutely convinced this was the wrong thing to do, but what choice did we have?

We all hated the man and his smelly guesthouse. Everyone was really worried about not having any ID, and Mehran wasn't speaking to me because he said it was my fault that we had given in.

But, as good as his word, the man dropped us at the coach station and gave us tickets for the bus to Tabriz.

It was another overnight journey, but this time I was too scared to get any sleep. When we finally arrived, it was into a much smaller bus station, and it was a relief not to be hassled by taxi touts on arrival.

A heavy-set Iranian Kurd approached us. "Qubat?"

By now we knew the ropes. "Yes."

He issued instructions to the others in a mixture of Farsi and a language I had never heard before, which I assumed to be Kurdish.

"It's time to go. We still have a long drive." He looked down at me and my little backpack. "Is that everything you have?"

I nodded.

"So then . . . *yallah*—let's be going."

I have no idea how, but six of us and our bags managed to squeeze into his Toyota. We rattled out of Tabriz and into the Iranian hinterland.

After an hour or so, the Toyota started to work harder as we curled up arid foothills. Our driver hummed along to loud pop music; he hadn't spoken a word to us since we'd set off. My legs had long since gone numb.

The road eventually straightened and I could see that, up ahead, it seemed to go right through the center of a large lake. Big black rocks covered in wire formed a barrier on either side of the road, and the traffic had begun to form a large queue.

I started to feel a sense of rising panic. Was this a border?

"What's happening?" I whispered to Abdul, who was sitting to my left.

He shrugged. He very rarely spoke.

The Kurd suddenly pulled over to the side of the road, got out, and opened the nearest passenger door. "*Yallah*. Get out. Let's move it."

On foot, we followed him, making our way up the side of the traffic. Up ahead I could see that the lake widened and stretched out into the horizon. The queues of cars were easing onto a large ferry.

The Kurd had gone over to some machines and had returned with a handful of tickets. He spoke in Farsi. "Act normally. Try to behave like tourists. Enjoy the view." He pointed to a gangplank where a handful of foot passengers were boarding. This was my first time on board a boat. And even though Afghanistan has lots of lakes, I had never seen one this big or this wide.

Most passengers stayed in their cars. There were no seats for foot passengers, and my legs swayed beneath me once we had found a place to stand, by the edge of the boat. The water was flat and placid but, as I looked over the edge into the deep, blue water I found myself having another attack of acute homesickness, and I wished I had someone to hold my hand and comfort me. But I didn't want the others to think I was a baby so I tried to act brave and nonchalant, as if I did this sort of thing every day.

The crossing took about fifteen minutes. As we approached the shoreline, the driver hissed for us to board a green minibus that was sitting in the center of the ferry. Behind its steering wheel sat a different man. We got in and, as the ferry came to a standstill, our new chauffeur drove off. About an hour later, in the middle of some quiet countryside, he pulled over behind a blue off-road vehicle.

"Please change cars."

They were the first and only words he spoke to us the whole time we were with him.

A third man drove us on. The road was good and as we traveled down an empty, winding road, the terrain began to change: trees dressed in October's autumn hues of gold and burnt orange lined our way. I was taking the first driver at his word, trying my best to enjoy the view.

"We should be there soon," the driver said.

It was nice of him to reassure us, except he failed to mention where "there" was. Mehran asked him where we were going but he didn't offer a reply. At some point, the driver took a call on his mobile phone. He spoke in short, hurried bursts of Farsi mixed with Kurdish, and so quickly that I could barely understand the Farsi. But I did notice that after the call his mood changed, and he became sweaty and nervous.

As a car approached on the other side of the road, our driver flashed it to stop and wound down his window. He spoke in Farsi to the other driver, so this time I could work out the gist of it: "Are they checking cars ahead?"

"No, I just drove straight through."

A few minutes later we cruised through an unguarded police checkpoint. That was a relief, considering we had no documents to show them.

"Allah was on our side today. They usually check the vehicles here." His relief was short-lived, however. Fifteen minutes later, as we wound uphill into a village, the engine cut dead.

"Shit. This isn't good."

We could see what he meant: there was a police station to the right, just a few hundred yards ahead.

"Stay inside and try to keep your heads low." He got out and lifted up the hood, fiddling around inside, then he came around to the driver side window. "It should be okay, but we need to jump-start it. All of you get out and push. Let's just be fast and get out of here."

We managed to get moving again just as two policemen ambled

down the road toward us. Thankfully, they barely paid attention as the engine barked into life and we piled into the car.

It was late afternoon when we turned into a cluster of low, concrete-block, and traditional stone buildings, surrounded by fruit trees and high fencing. Dozens of cars were lined nose-to-tail in front of a dilapidated cowshed. Waiting in the yard was an imposing-looking man.

"I am Black Wolf," he said, shaking our hands as we got out of the vehicle.

His eyes were as dark as the color of his name, but they seemed kinder than the name suggested. He shook down a large gold watch from beneath the sleeve of his white tunic, making a show of checking the time.

"Welcome to my house." He gesticulated with a sweeping arm. "I apologize for your inconvenience with the different vehicles. But you are here now." He smiled, showing near-perfect white teeth. "Welcome to our Kurdish heartland. I will show you where to sleep, and bring you some water."

He spoke beautifully articulated, crystal-clear Farsi, even though he, like the driver, was obviously Kurdish.

"Ignore these men," he continued, waving his arm at three men busily syphoning fuel from various vehicles' petrol tanks into battered orange fuel drums. "They will not bother you, and it is none of your business, anyway."

Whatever Black Wolf was up to, I got the impression business was good.

"Come, gentlemen," he said. "Follow me."

We followed him past a pile of car parts and scrap metal, between two further stone buildings, and toward a large, very modern-looking house. We entered beneath a smart-looking porch. Black Wolf pointed to a door just to our right, and I noticed a large diamond and emerald ring glinting on his finger. "Here," he said. "This is your room."

It was well furnished, with traditional mattresses around the sides and Iranian carpets on the floor.

Through the net curtains of the window, I could see trees heavy with apples, oranges, and pears—the likes of which I hadn't eaten since I had left home. My stomach went into a frenzy of rumbling.

"You can help yourselves to fruit anytime," Black Wolf said. "But be quiet and discreet." Then he smiled and ruffled my hair. "Don't look so worried, boy. No one is going to hurt you. You are going to rest here for a few days until I get word it's time to move on." He looked up at the rest of the group. "The next part of the journey is hard. You will need strong legs, so I expect you all to behave with good manners and sleep as much as you can. If you see that I have other guests, please stay in your room. You definitely do not want to make any trouble for me."

With that, he turned on his heel, speaking over his shoulder. "Someone will bring you some food shortly. There is a tap and a toilet around the side. I suggest you make yourselves comfortable."

Too exhausted to talk, we flopped down onto the mattresses.

I WAS SHAKEN AWAKE.

"Hey, Gulwali." It was Mehran. "Wake up, man. It's food time."

I was confused by the half-light.

"It's almost dark. You've been snoring for hours."

A lanky, teenage boy was in the room. He placed a large dish of rice, a steaming roast chicken, a pot of lentil stew, and a traditional Kurdish salad of olives, cheese, cucumber, tomatoes, lemon, and cabbage on a cloth in the center of the room. It was a feast.

Black Wolf strode in carrying a pile of plates. "This is Rizgar, my brother's son." He nodded at the boy. "Come on, then. Take a plate—get eating."

He smiled in my direction. "Don't delay, little boy. Get eating before the fat one beats you to it."

Shah glowered at him as we all laughed.

As six hungry people sat eating cross-legged in silence, I found myself staring out of the window into a star-filled sky. The stars shone almost as brightly as they did in the mountains of my childhood. It had been just over a week since I had been ripped away from my family and everything I had known, but the pain of each hour had made it feel like months. After the meal, when we said our late evening prayers, I asked Allah to keep my family safe.

I WOKE VERY EARLY THE NEXT MORNING. IT WAS STILL DARK. The others were fast asleep, so I stepped around them quietly and went out to visit a beautiful horse I had spotted in a field next to the orchard the previous evening. I took a pink apple from a tree and held it over the fence. I clicked my tongue against the roof of my mouth and a large shadow in the distance moved closer.

The mare was a beautiful chestnut with a shiny coat that glowed even in the pale light. Her muzzle twitched as she scented my offering, and her large white teeth crunched up the apple in two easy bites.

It felt peaceful being alone with this graceful animal, and I couldn't help think of my early years living with my grandparents in the mountains. How could I have had any idea then of the pain and suffering that was to come to me and my family?

The strains of the dawn call to prayer floated across the trees as a softly pink light began to melt the clouds. I hurried back through the orchard, back toward the main house.

"Little boy."

The voice felt loud in the still morning air.

"Hey, little boy."

Black Wolf loomed over a low wall that marked the entrance to the back of the house. "Are you coming to pray?"

"Where?" I asked, unsure where I should go.

He waved a hand, his jeweled ring glinting in the dawn, beckoning me toward him. "Come, boy. Come inside."

I stepped through the back door into the inner sanctum of the family quarters. A group of figures stood in the middle of the central courtyard, facing toward Mecca.

"Little boy, this is my brother and my cousin. My nephew you already know." He nodded to me, and then addressed everyone. "Let's pray, brothers."

I joined them as they fell on their knees. I thanked Allah for a merciful journey and prayed for his blessing on my mother. I asked him to keep my brother safe until we were together again. And I thanked him that Black Wolf was a kind man.

When we finished I felt much better—very calm and comforted.

Black Wolf looked over at me. "Will you take some breakfast with us?"

I beamed. "Thank you." It was a great act of generosity by Black Wolf to allow a male who wasn't a direct relative to meet his family. I was nervous. This was the first time I'd been in the company of Kurdish people.

I knew from geography classes at school that the Kurds are the largest ethnicity in the world without a country of their own. They historically inhabit the border regions of Iran, Turkey, Syria, and Iraq, but they don't recognize the boundaries between these countries. Like the Pashtuns, they also have a very strong sense of identity.

Perhaps I should have been more careful, but Black Wolf had a relaxed manner and an easy charm, and I was fascinated by him. The way he moved had a confidence about it—as if he knew with absolute certainty he was the lord of all he surveyed. He reminded me of the tenth-century merchants I had read about in history books—the men who made this region famous for its rich trade routes.

If I still had any misgivings, they evaporated the moment his wife and her sisters-in-law caught sight of me. For a moment I was taken aback at his wife's beauty: she wore a loose scarf around her neck, allowing her long black hair to escape onto her shoulders, but the little boy who had taken family honor so to heart back at home became outraged. I wanted to tell her to cover her hair more carefully. "Oh, isn't he adorable?" Black Wolf's wife gushed, tousling my hair in a way intended to embarrass and delight me in equal measure. "How cute." Her teasing continued, even as the men sat to eat. We ate flat bread, rich with olive oil, dunked in smoky baba ghanoush and some of the dal I recognized from the previous evening's meal. We finished with delicate glasses of sweet, black tea.

Black Wolf sat back and lit a cigarette, picking a fleck of breakfast from his teeth. Tendrils of smoke oozed from his nostrils. "So," he said, cocking an eyebrow at me. "Are you recovered from your journey so far?"

"Yes, thank you." Then, to my horror, I suddenly felt tears spring to my eyes—perhaps because his family reminded me so much of my own. I chastised myself for being so silly and tried to hold them back, but it was too late.

"I wish I could say I hadn't seen many children like you on this path, but I have. Too, too many. It's a sad world."

My tears flowed unchecked now, and when his wife dabbed at my eyes with a tissue, the act of maternal kindness was just too much. I grabbed the tissue from her and looked at the floor as I struggled to find the right words in Farsi to make myself understood: "My mother sent us away. I don't know where my brother is." At that, a series of big, gulping sobs escaped me.

Black Wolf and his wife looked at each other and shook their heads sadly.

"This is unfortunate. But I am sure you will find him again. You have a long way to go and will no doubt cross paths with many unscrupulous men. Take my advice, little boy—do not trust these people, these so-called people smugglers."

"But aren't you a people smuggler?" I blurted out, immediately regretting the way it sounded.

Black Wolf bristled for a moment, taking another pull on his cigarette. "Me? You do me an injustice, boy—and under my own roof, too."

His family laughed at my obvious discomfort.

He waved a hand for dramatic affect. "I am a mere facilitator. My business interests lie in other directions."

His brother, cousin, and nephew laughed again. His wife gave me a stern look.

"No, I am merely a man who has a well-located farm where people like you can stay until the smugglers are ready to move you on. I provide a service, and offer a level of discretion these men find comforting."

With that, he sent another cloud of smoke billowing up to the yellowed ceiling.

When I rejoined my friends, I wasn't their favorite person.

"Gulwali, you idiot, where the hell have you been?" Mehran did not look happy. "You missed prayers."

Even Baryalai raised his voice. "I told you to stay with the group. Do you think we need the drama of worrying about you as well as ourselves?"

I tried to joke my way out of the situation. "What? Did you guys eat breakfast without waiting for me?"

Even Abdul said a rare word: "Gulwali, you haven't answered their questions."

I was about to tell them, but then something stopped me. I had a sense that, however nice they were, they'd resent the fact I'd just been given special treatment.

"A horse," I said. "I was playing with the horse."

FOR THE NEXT FOUR DAYS, WE SETTLED INTO A FAIRLY RElaxed routine of eating, sleeping, praying, and playing cards. Or rather, the others played cards—no one offered to teach me, and I felt too embarrassed to ask.

There were a couple of Kurdish storybooks on a shelf in the room. Like Pashtu, Dari, and Farsi, the Kurdish language originates from Arabic; and, like most Afghan kids, I'd been taught to read the Quran in Arabic from a very young age. So I figured if I tried, I might be able to make out a few words.

I spent hours poring over the book trying to make out a few meanings.

On the fifth day, Black Wolf came to see us in the morning, clearly troubled by something. "I can't make contact with the agent in Turkey. He's supposed to be meeting you there."

Abdul was the first to speak: "So, what does that mean for us?"

"I can't make my arrangements to get you safely over the border until he makes contact. This has never happened before. I've only been paid to have you here for three days."

I was beginning to learn that everything came down to money with these people.

Black Wolf must have read my thoughts because he looked straight at me. "I'm not a greedy man. You can continue to stay here. And you will be taken care of. But, understand, this is causing me inconvenience. I have various meetings with business associates over the next few days. I need you to stay out of sight. After breakfast, you will stay in the furthest back orchards until it is dark. One of my men will bring your food there, and you will not return to this room until you are given instructions to do so."

We sat listening in silence. I supposed it could have been worse.

For the first day it wasn't so bad. The late autumn sun still carried warmth, the trees offered dappled shade, and Black Wolf had laid out a large, woven-plastic mat for us to sleep on. When my stomach started to rumble in the afternoon, I plucked fruit from the trees.

Black Wolf entrusted me to collect our food from the women in the kitchen. I didn't mind in the least—it felt special to be trusted by the family, and the women insisted on doting on me in a way that made my insides bloom with butterflies.

By now we'd been there ten days, and our relaxed routine was beginning to feel like a prison. We wanted to move on as much as Black Wolf clearly wanted us out of there. We couldn't stay here forever, especially not without a passport. I had no option but to continue on to Turkey and the ultimate destination of somewhere in Europe. That was as much as I knew. And of course Hazrat was out there somewhere. I could only find him if I kept moving. The situation only got worse with the arrival of another five refugees. They were also Afghans. At nighttime the room was already squashed with the six of us; with five more bodies it became impossible to sleep without having to lie on someone's legs or end up with a foot in your face. I didn't like the newcomers, while my friends were growing more sullen and angry by the day. I tried to avoid everyone by going to the furthest reaches of the orchard where I could be alone.

One evening, I was out there on my way to see the horse. I had visited her daily. Patting her soft auburn nose made me feel calm. As she stared at me with her mournful brown eyes, I felt as though she could see right into me, sensing my fear. That was when Baryalai came running through the heavy dusk. "Let's go. Get your stuff. We're going."

"We're really going?" I asked, not quite believing my ears.

"What did I just say? Now, get your stuff, little man."

I paused, wanting to go see my horse before we left.

"Come on, get moving."

Baryalai was already running back toward the house. I looked back but I couldn't see her. I followed him.

The whole thing was like a military operation: within minutes, all eleven of us were crammed into the back of a high-sided pickup truck. There was hardly time to shove my things back into my backpack, let alone say good-bye to his wife.

"Stay down and shut up," Black Wolf said. "And don't move for anything." As the truck rattled down the road, I looked back at the orchard, still hoping to see the mare among the trees there heavy with fruit, but I couldn't. The field was empty.

CHAPTER 7

FOR MOST OF THE JOURNEY WE DROVE WITH THE HEAD-lights off. The sky was carpeted with stars, giving enough light to see the splintered rocks that studded the dim contours of the surrounding hills. We were heading into a terrifying blackness along mountain roads, toward the crossing into Turkey.

It was cold on the back of the truck—the slipstream rushed down my back and made me shiver. But the air smelled sweet with the scent of wild grasses and herbs.

Before long, the truck slowed. Black Wolf called back to us softly. "Keep calm. We are coming to the police checkpoint."

The eleven of us were squashed together tightly, but I huddled myself down to be even smaller.

The driver's window squeaked indignantly as Black Wolf wound it down. Then, the scorching beam of a flashlight swept into the truck.

I can only guess that this policeman—whoever he was—had been paid off by Black Wolf in advance: there would have been no mistaking the eleven cowering bodies in the back. But he shook Black Wolf's hand and simply acted as if we weren't there.

Black Wolf wrestled with the gear lever, before over-revving the engine and lurching us forward with a roar of tires spinning on gravel.

I must have nodded off because my next memory is of waking up with the truck bouncing violently over broken ground.

"We are here."

Waiting for us were the two horsemen who had been sent on ahead. As we got out of the pickup, I looked curiously at the oil drums hanging on each side of their saddles.

One turned his horse toward the mountains. "Follow me. Stay close. And don't fall back."

As we prepared to follow them, I turned to say good-bye to Black Wolf. To my disappointment he was already in his truck, pulling away without even looking back at us. That hurt.

We started walking, the second horseman bringing up the rear. I could make out the shapes of other horses and people walking parallel to us in the distance. Coming from the other direction, we passed a group of sad-faced donkeys pulling rickety carts piled high with goods: boxes, crates, sacks. As we continued further up the slopes, we saw yet more people going back and forth, some with horses slipping and straining beneath heavy loads.

We stopped to drink from a little brook. The night air was so still, the mountains acted like an echo chamber. The unmistakable sounds of people singing folk songs drifted over to us.

I was amazed by the scene. Was this a border? It reminded me of the landscape I'd traveled to reach Waziristan. I had thought of Black Wolf as a tenth-century merchant, and this sight was even more fittingly incredible. It was like an ancient caravan of traders.

A familiar voice rang out: "It is as if King Darius himself is assembling his army tonight, yes?"

Black Wolf's lanky nephew, Rizgar, was walking toward us, leading a horse. It struggled beneath bulging saddle bags. Whatever was inside them, I assumed, was in exchange for the oil Black Wolf was so clearly smuggling into Turkey.

"All of humanity is here," he said, as he high-fived me. "We have cheap labor, cheap fuel"—he knocked theatrically on one of the oil drums—"and let's not forget cheap thrills—how much of your famous Afghan opium can a camel carry, do you think? This, my friends, is where the real business of the capitalist world is done."

And with that he disappeared back down the hill, seemingly swallowed by the darkness.

WE WALKED THE WHOLE NIGHT. I'D NEVER WALKED THAT FAR or on such rough terrain. I was exhausted. The ordeal was made worse by not knowing how much longer we'd be walking for, or where exactly we were going. I knew we were walking toward Turkey, but no more than that.

Just before dawn, we clambered across some stones that bridged a little riverbed, and rounded a bend. My stomach went into little fits as heavenly wafts of chargrilled meat floated through the air, while the strained notes of Turkish pop music competed with a growing hubbub of voices and the sound of braying donkeys.

On a small floodplain at the bottom of a mountain was a makeshift bazaar. It was alive with hundreds of people and their packhorses, while battered trucks honked their way through the crowds. Everything was being sold in bulk, like an open-air warehouse: cooking oil in drums, flour by the sack, raw wool, hessian bags of corn and beans. Some meat was for sale—mutton, mostly—and all still on the hoof. Men haggled, waving fists of grubby notes at each other with an intensity that suggested a punch might be thrown at any moment. A woman was busy loading a goat into a truck as her husband yelled instructions at her.

Our horse-riding guides led us to a small guesthouse that was built into a stone cliff. They told us to wait inside, then left us.

A Kurdish woman with a face as crinkled as a walnut shell directed customers with the no-nonsense manner of my beloved grandmother. She too spoke the local Kurdish dialect, and she kept trying to talk to us even though it was obvious none of us could understand a word. Eventually she worked this out and, using sign language, told us to follow her. We were delighted when she showed us a small bathroom where we could wash.

When we came back, she had produced a large platter of bread, olives, and a type of feta cheese. Happy to be clean, I set about the food. I had never tasted such sweet bread, while the olives and cheese were salty and rich. I could feel my body thanking my mouth, expanding in gratitude.

After we'd eaten, our guides came back and told us to follow them outside. The horsemen walked us a little way up a hill, then simply trotted off. "We leave you here," one called over his shoulder. "God be with you."

We stood there perplexed, not knowing what to do next. Thankfully, a small red pickup truck was waiting nearby, and the driver gestured us over: "You. Qubat people. Get in please."

The back of the vehicle was packed with rolled-up carpets—he tried to squash the eleven of us in too. I was thinking how stupid it was. With such a heavy load, I couldn't see how it would go anywhere. Somehow we crammed inside, and the driver put blankets over our head. The pickup grumbled along slowly and painfully.

From my vantage point, I could see some of the road, and my heart was in my mouth as we passed a roundabout. A man in uniform, I think a local militia, was directing traffic. He had a rocket-propelled grenade launcher strapped to his back.

It was as we began trundling down a rocky valley that we, unsurprisingly, broke down. We all got out while the driver tried to fix it. There was so much dust on the road it felt like snow on my feet. Even in the desert landscape of my home I hadn't seen anything quite like this.

We suddenly became aware of the sound of a car approaching. Further along the valley, about a kilometer away, we could see a vehicle snaking toward us.

"It's a police car. What shall we do?" Faizal yelled in panic. "We have to get out of here."

We tried to push the pickup, but the carpets inside weighted it down. We pushed and pushed with all our might until finally it spluttered into life. Trying to squeeze ourselves in a second time, this time with real urgency, was even harder.

After driving for a few more miles, we arrived at a small guesthouse. There was a group of six other Afghans there already. They didn't speak to us much other than to tell us they were all from the province of Kandahar.

All I really remember about that stay was that my feet were so hot from being half-smothered by blankets, that I almost expected steam to come off them when I rinsed them under the outside tap. I was so grateful for the water—the way it cooled my body and helped settle my mind. As I sank into sleep, I didn't care that the tiny room I was in was cramped or that the mattresses were alive with fleas.

Less than thirty minutes later, however, we were rudely awoken by the guesthouse owner. He was carrying a handful of disposable razors, while with his other hand he gestured to his chin, making a shaving motion.

The men were on their feet in a flash.

"You can't be serious," said one of the guys from the other group, pulling at his thick gray beard to emphasize its length. Then he pointed to me. "I have grown this since I was his age." The manager didn't understand his Pashtu; he just continued to make the same shaving motion.

The old man turned to the rest of us. "He cannot be serious. What if we are caught and I am sent home? How can I stand in front of my family with a shaven face?"

I was much too young to shave, although I could sympathize. All of my male relatives were proud of their facial hair. A long and full beard had not only been the law in Taliban-controlled Afghanistan, for many it is a key aspect of what it means to be an Islamic man.

They soon returned from the bathroom with faces like plucked chickens.

"Don't you dare laugh," said Faizal, holding his fist to my nose. Abdul couldn't stop giggling. "I'm serious. One more laugh and I'll cave your heads in with a brick."

A turbaned old man sat in silence, running his fingers over his smooth chin. Silent tears dripped down his face.

There was nothing I could do for him. I decided to go back to sleep.

THE NEXT DAY, A FACE PEERED AROUND THE DOOR AND SPOKE to us in a mixture of Kurdish, Arabic, and Farsi: "Good morning. I'm glad you arrived safely." He looked around the room. "Ah, that's good," he exclaimed. "Before, they were only Afghan goat herders—now, they are taxi drivers."

I seethed inside. Not only was he disrespecting the elders, he was making offensive comments about shepherds. My own grandfather was one so I took this personally.

"Now," he continued, holding out a battered Polaroid camera. "Let's capture that beauty for the ages. You need new documents."

The rude camera operator told us he was an agent working for Qubat, and that he would be responsible for the next leg of our journey. He explained we were in the first small Turkish town in the Kurdish region, literally just over the border.

According to him, there were two different ways to a place called Van, the nearest city. Neither option sounded very appealing. The first was to walk—it was less risky but much longer and more difficult. The second option was to go by car. This was obviously much easier—but the risk of arrest was much higher. If we wanted to take the second option, which he intended for us to do, that required new fake passports. Baryalai shook his head in frustration at this. "We had some but you people took them off us in Iran." It took another day to get the documents made but it was, we all agreed, worth it. No one had wanted to walk.

When it was finally time to go, I was ordered into a red Toyota Corolla along with Baryalai, Mehran, and Abdul. I made sure I held Baryalai's hand. The others, including Faizal and Shah, got into different vehicles, all of which drove off in different directions.

"I have a feeling that is the last we'll see of them," Baryalai said sadly.

We drove a few miles around the town before swapping to yet another car. The agent had told us secrecy was necessary because in a small town like this one, outsiders were easy to spot and people were inclined to talk.

The driver of the second car was a bit cocky. I didn't feel comfortable the moment we got in.

"My car is fast," he said, stroking the carpeted dashboard. "We will arrive soon, *Insh'Allah*."

By now I had picked up a few words of Kurdish and could make out most of what he said. While invoking the will of Allah was normal in such conversations, his sales patter was less than orthodox: "Pay and pray," he sung to us, as if it were a radio jingle. "That's the way to make your stay,"

It didn't really make sense to me. But then we were only paying for his car, not his conversation.

The scenery was wild and beautiful. Jagged ridgelines of the steep hills plunged into wooded pockets in the narrow valleys and gorges. I was very pleased we weren't walking.

The winding road had a constant military presence—we passed a long convoy of armored vehicles and earth-moving machinery atop trucks, all heading toward the border. Turkish infantry marched along the road, too. I recognized the uniforms from home, where Turkish troops were part of the NATO mission.

As we rounded a bend, a manned roadblock suddenly appeared. Cars were being pulled over to the side and searched by soldiers.

"Say nothing," the driver ordered, staring at us in the rearview mirror.

An officer stood in the middle of the road, directing traffic toward his waiting men. An AK–47 hung around his neck, its muzzle tracking at windscreen height. He signaled to us to pull over. The car slowed to a rest in front of a sandbagged machine-gun position. Our driver lowered his window and volunteered our passports, speaking in Turkish. My heart thundered in my ears. I knew that if the officer spoke to me, then the game would be up. I put my hand over my mouth to stop myself blurting out something in fear. The officer took his time, reading each passport and staring into the car. We hadn't seen the new passports until this moment. I was surprised to see they weren't Turkish, as we had assumed they might be, but the familiar Afghan blue. Why would they give us Afghan passports again, when they had taken away from us the ones we already had in Iran? And how would an Afghan passport work here? None of it made sense.

More exchanges.

And then the soldier handed the passports back. He smiled into the backseat. "Enjoy your holiday."

"Praise to Allah," the driver shouted with a laugh as we gathered speed away from the roadblock. He held out one of the passports and pointed at what looked like an entry or visa stamp. "Potato. We make with potato. It's visa *aloo*."

He thought this was hilarious.

But it seemed the potato stamp had fooled the soldier into believing we'd entered Turkey legally.

"Two more checkpoints," said the driver, happily. And then he put out his hand, gesturing that he wanted money. "Pay, please pay."

"The agent has already paid," Baryalai protested.

"You've paid me? No. Not me."

"We paid Qubat. He paid the agent. You've been paid."

"Really? Maybe you think of my cousin? People say we look the same."

We all knew this guy was just trying to scam us, but we were resigned to the fact we had to give him something.

One by one we pulled a few dollars from our pockets. I had spent most of my Iranian currency on snacks and drinks in Iran, but I still had the $200 my mother had given me.

I took a couple of my smallest notes from my wallet—the most rumpled and filthy ones—and added them to the little stack on the driver's elbow rest.

"Yes," said the Kurd. "Pray and pay. Very good."

He put the passports firmly back into his glove box. Clearly we didn't get to keep these ones either.

I felt nauseous. Everybody was out for money in this game, and we were very easy targets.

AFTER ABOUT HALF A DAY WE ARRIVED AT VAN. WE DROVE through a series of dilapidated neighborhoods before pulling into a complex of abandoned workshops. We all climbed out and waited in the shade as the driver approached a tall, thin man in a Western-style pinstripe suit.

They shook hands and had a short conversation before the driver got back in his car and departed. "Pray and pay," he shouted, by way of farewell.

The tall man looked very Turkish, yet he spoke to us in Iranian Farsi. "I am Malik. This is my operation. In a few minutes, one of my men will transfer you to a minibus, which will take you to a house in the suburbs. There, you will be able to rest before your journey continues."

I thought he looked very impressive. He carried a leather briefcase and exuded the air of a man who was very much the boss. He was like a Turkish version of Black Wolf. I started to imagine how nice it would be to stay in a decent place again.

I got very excited when we saw the house. It was large, sitting behind a high wall, with many shady trees in the garden. It looked beautiful. I was already planning to lie beneath a tree on the cool grass.

Our driver pulled up right outside. "Go quickly—but don't run."

As I stepped through the small visitor's door on the main gate, however, my heart sank. There were two burly guys standing there, neither of whom looked friendly.

"Come. Around the back."

The two guys manhandled us, pushing us along a narrow passage that ran between the house and the high brick wall. At the rear of the property stood a little shack.

Baryalai was the first to speak.

"It's a bloody chicken coop."

CHAPTER 8

"WELCOME TO HELL," A RED-HAIRED MAN WITH A STRAGgly beard said as we entered the coop.

A few mattresses covered the floor, but there was not enough room for all of us to lie down. There was barely even room to sit. Eleven men stood around us, dead-eyed and disgusting with filth. The four of us squashed ourselves in. A tiny, plastic-covered window gave a blurry view of the trees in the garden.

Previous residents had written on the broken door, which swung despondently, hanging off its hinges. Some had chosen to write their names and the dates they had been there, using the Afghan calendar. All the entries were written in either Pashtu or Dari.

Hamid Shah. Herat. 1384.

Khalid Kakar. Kabul.

Others offered advice or warnings:

Trust no one.

Do not pay ferry man.

Life is cheap and so are smugglers.

Malik is a bastard.

In the coming days, I learned the truths on that door.

A scruffy, bald man in a dressing gown brought us some food: boiled rice and Turkish bread. The rice was dry and the bread looked stale;

nor was there nearly enough to go around. He didn't speak a word as he roughly plonked the dishes on the doorstep.

After he left, the others let rip.

"This bastard has no manners."

"He treats us like animals. Get used to it."

"He's a filthy drunk."

They said they'd been there for three weeks, yet Malik had promised them it was only supposed to be overnight. They were waiting to go to Istanbul, on the next leg of the journey. I assumed from this that Istanbul was to be our next destination too.

We were not allowed outside that chicken coop. We had a single pipe that only dribbled water for half an hour twice a day; we never knew what times it would be turned on, so when it did splutter to life we would scramble to wash our faces and rub a toothbrush through our stale mouths.

A lot of time was spent trying to control our bodily functions. Just once every twenty-four hours we were taken in groups of three to use a stinking toilet. There was never toilet paper, and the tap for washing yourself often didn't work, forcing desperate men to wipe themselves clean on the streaked walls. It wasn't long before people started to get sick. I don't know what was worse—the horror of watching someone vomit in such close confines, or the embarrassment of the man who soils himself in front of his travel companions.

For the entire time we were there, we got very little food—a few mouthfuls of rice each per day, if we were lucky. When we complained, the bald man said he'd bring us a roast chicken if we paid him. He charged the equivalent of $5. I was shocked—my mother could have fed our family for days with that money. But I was starving, so I spent some of my precious money and ate my little share of the chicken greedily.

I did my best to cheer everyone up by trying to tell jokes or reminding them of one of my favorite Pashtu sayings: "If the heart is big enough, the space is never too small." I can't say my efforts really worked.

From our few toilet breaks, we worked out that the bald man lived with his elderly mother. Occasionally, there were a few kids about. I could see, too, why the others had called the man a drunk: he stank of stale alcohol and was always wearing his dressing gown. This made me dislike him even more—I totally disapproved of any Muslim drinking alcohol. His speech was permanently mumbled and he seemed strangely blind to the appalling situation we were in. This man was capable of such cruelty, but the odd thing was he seemed numb to what he was doing.

Fortunately for me, the elderly mother noticed how young I was. She did not shy from showing her affections, and I had already learned at Black Wolf's compound the benefits of ingratiating myself with the women of the house.

We had no means of communication, however, so without a common language we resorted to a kind of sign language.

I pointed at myself. "I am Gulwali."

"Gulwali," she repeated, struggling with my accent. Then, "Gu-wal-i," she said again, breaking it into syllables.

I nodded enthusiastically.

She smiled a toothy smile and pointed at her chest. "Ma-ree-ammm. Ma-ree-ammm."

"Mariam?"

She seemed pleased with my progress. Then she pointed at the bald man. "My son." At that, she rolled her eyes then pointed to her temple and twirled her finger in a circular motion, as if to show he was mad.

I sensed an ally. "I am twelve years old," I said, explaining it by pointing at my chest then flashing first ten, then two fingers.

Her face erupted with concern and surprise. She pointed at the dilapidated shed and waved her index finger horizontally to show the universal sign of matriarchal displeasure. Then she drew her finger toward the back door of the house and pointed to a small patch of carpet in the family living room. She stabbed her finger at the floor, then placed her palms together and rested her cheek upon them.

"You want me to sleep in there?" I asked, repeating the gestures back at her.

She nodded enthusiastically.

I broke into a huge smile and, without a moment's thought, gave her a huge hug. "Thank you."

I will never forget the kindness and gentleness that old lady showed me. In addition to the safe place to sleep away from the hell of the chicken coop, she also snuck me into her kitchen and forced bowls of rice and vegetables, Turkish bread and steaming sweet tea into my hands.

Those left in the coop were furious that I was allowed to sleep inside; it made no difference to them I was a child. I'm sure they suspected I was getting fed well too. They thought I was manipulating the family to get special privileges. Maybe I was. But would those men have done any differently in my situation? I was a little boy, and I was fast learning it was the only card I had to play.

I think the truth was that we were all so desperate that we quickly came to resent anybody who had something we did not—the extra mouthful of water, a tiny bit more floor space, a filthy pillow, or a few grains of rice. Our humanity was slipping away—being stolen away. Perhaps that was the real price of this journey.

The resentment grew as the situation inside the chicken coop deteriorated. We had been there more than a week and some of the men were beginning to reach breaking point. Starved, thirsty, and packed into filthy conditions, they became increasingly desperate. At least I was able to use my position inside the house to help my friends a little bit, sneaking in extra food and water.

Early one morning, a man turned up in the yard carrying several lengths of heavy polythene sheeting. He balanced on an old chair and began to cast the sheeting onto the roof of the shed, like a fisherman casting his net. Satisfied with how he had laid it, he then lobbed half a dozen broken bricks on top to hold it in place.

The *thump-thump-thump* on the flimsy structure woke several of the occupants inside.

The bald man, who had come down to watch him work, grunted and pointed, directing their gaze to the new roof covering. The builder wore the half-smile of a man who has done a good job. "Now that I have fixed your dog kennel, I will be able to get another twenty or thirty dogs to join you, no problem."

It was too much for the men to hear. One of them lunged forward, grabbing a brick from the pile on the floor. "Where is that liar, Malik?" He grabbed our drunken host by the wrist, turning the half-brick toward his face. "Maybe I should knock some sense into you instead?"

Two other men surrounded the builder.

There was a short pause and then, without warning, the builder suddenly crouched into a ball and flew at them, sending them flying onto their backs.

Everyone else began kicking and punching.

It was madness. We would be discovered, I was sure of it, so I shouted at the top of my lungs: "Stop it. Stop it."

"Stay out of it, boy, or you'll get a slap too."

I backed away just as two other men—the ones in charge of the main gate—ran into the yard. They both had sticks, which they used to beat my friends until they lay whimpering on the ground. They were about to come for me but Mariam pulled my face into her bosom and wagged her finger at them.

They shouted something back at her in Turkish, waved their sticks once more, then took out cigarettes, lit them, and swaggered back to their guard post.

Mariam took me inside and insisted I drink some tea.

A little later, one of the guards came and thrust a mobile phone into my hands.

"*Salaam,* Gulwali. This is Malik."

I held the phone in silence.

He continued to elaborate, in Farsi: "Gulwali. I heard about what happened this morning. And I hear you have all been complaining. This is not good. I want you to talk to the others and inform everyone that I am trying to get you out of there as soon as I can."

"Okay," I said, unsure about why I needed to be the messenger.

"I am arranging a luxury coach for the next leg of your journey. Lots of room, air-conditioning, comfortable seats. But such luxury does not come easy here—I need a little more time to organize this. So tell everyone to be patient, because otherwise they will have to travel in the back of a smelly truck."

"Okay," I repeated. "I will tell them."

"Good boy. And don't forget—it will be a luxury bus. Air-conditioning and all that."

I soon realized why Malik hadn't wanted to deliver his news in person: the men glared at me with pure hatred.

"And you believed him? You stupid, foolish boy."

"I don't care how we travel. I just want to get out of this stinking, shit-filled cage. But I suppose you didn't tell him that?"

"And besides, what do you care? Tucked up inside with your foster mother dropping morsels down your scheming little throat. I should choke you like a chicken."

"Leave the boy alone," said Baryalai. "Can't you see he's no more to blame for this than ourselves? Until we reach our destination, we are at the mercy of these people. They own us, and can treat us as they will. You'd better get used to that."

"Easy for you to say," snapped one of the eleven. "You've only been here one week. We've been here for almost a month now. And it's more than half a year since I last saw my wife and children."

Later that day, Malik himself strolled into the yard. He was all smiles and greasy charisma. "And how has the food been, boys? Has the man been feeding you properly? You look as though you've put on weight."

At that, the men couldn't contain themselves a moment longer. They railed at Malik, listing their grievances one by one, careful to highlight the near-starvation conditions we had been forced to endure.

"I am as shocked as you are," exclaimed Malik, holding his hand to his heart. He loosed a storm of abuse at the bald man, who had come out to the yard with him. He was clearly drunk again, and you could see he was confused. I found it hard to believe that Malik didn't know exactly what was going on—he struck me as the kind of man who kept a close eye on his business.

I concluded that Malik was shifting the blame. He clearly liked to protect his self-image of a successful, maybe even legitimate, businessman. Any failings were out of his hands—he was trying to do right by us. But even in my child's mind I knew our misery paid for those expensive suits.

As he left, he gave a curt promise: "The situation is urgent and I will deal with it accordingly."

The following afternoon we were ordered out of the property.

CHAPTER 9

OUR DEPARTURE WAS SO SUDDEN I DIDN'T EVEN HAVE time to say a proper good-bye to Mariam.

I owe her so much—her kindness may have literally saved my life, as I'm not sure I could have survived in the shed with the men. I also felt sorry for her: she clearly wasn't pleased with having a bunch of undocumented refugees in her backyard, but with a useless, alcoholic son in charge, what power did she have to say no?

We were driven to the outskirts of the city. By now, I understood fully that behind all the noble talk, the only thing that mattered to any of these people was money. This was a business, where profit was the only motive and people were of value only for as long as they could keep on paying. But there is more than one way to make money out of refugees—as I was about to discover when we pulled up outside a large, ramshackle hostel building, just off a ring road.

"I think I have died and gone to heaven," joked Mehran. He stood, slack-jawed, staring at three girls bending over tables collecting dirty plastic plates and cups.

I had never seen a blond woman before. It shocked me—not so much the color of her hair, but rather that she made so little effort to keep herself covered. She was wearing a skirt that barely reached her knees, and a tight T-shirt that left nothing to the imagination. To me she was as good as naked.

"I doubt these women are virgins," snorted Abdul. He looked at the women in a way I thought was very bad, and I shot him an admonishing look.

As we walked down a long corridor inside, I could see various rooms off to the side. There must have been over a hundred people staying there—the cacophony of different languages filled the air.

Dozens, perhaps more than fifty people, sat on the floor of one of the largest rooms. Mostly men and boys like ourselves, they ate with a concentration that comes from constant hunger.

A young Persian woman with eyes the color of Turkish coffee and an elegant oval face called out to us in Farsi: "Come. Sit and eat." There was a sadness to her, and a heavy layer of makeup did little to hide the swelling under her eye. The bruising on her upper arm also spoke of violence.

"These women are filthy whores," said a man, loudly, as if condemning their morality might make him a better human.

"Then don't eat their food," said Baryalai. He smiled graciously at a woman as she handed him a heaped plate of delicious-looking red peppers stuffed with vegetables and rice. She was a little older than the others—fairer than most, too.

"She's European, I am sure of it," he said, as if he could read my mind.

"They should cover their hair," I said. "Are they not ashamed to disgrace their families by dressing in such a way?" For me, modesty and honor were two sides of the same coin.

As we crossed our legs and began shoveling in food—I hadn't eaten properly in weeks—Abdul looked at me thoughtfully. "Gulwali," he said, through a mouthful of food. "How did you come to be here?"

"You know this. My mother sent me and my brother. We had to leave."

"Yet you think these girls are here by their own choice?"

"No, I'm just saying they should cover themselves. Their families—"

"We're a long way from home now, Gulwali. This is not Afghanistan. None of us have families. Are these women not showing you kindness?"

"Yes."

"But still you would condemn them as immoral?"

Abdul didn't speak much, but when he did, I have to admit it was usually with wisdom.

Mehran sat himself down next to us. "I can't decide which one to marry first. The Persian is kind and beautiful." He took a mouthful and continued: "She is sweet and a little shy. She would make a good wife, I think."

Baryalai looked at him. "I don't think these women are open to marriage. Or, more likely, whoever owns them has plans for them that don't involve marrying a penniless Afghan sneaking his way to Europe."

I smiled. "You sound like my grandmother."

We all laughed.

"Clearly a wise woman."

As I ate, I remembered how the sheep used to know the sound of my whistle. I had been their master. Now we were the sheep, and our master's voice was currently the lilting baritone of Malik and his purring lies. Was he also the master to these women? Did the business of prostitution also pay for his smart suits and leather briefcase? I couldn't have hated him more at that moment if I had tried; he was the epitome of immorality to me. This was everything my family and my culture stood against.

We spent twelve hours there—long enough to realize that there were other women there. Some were Africans, one who looked definitely Middle Eastern, a Chinese woman, and a couple more who looked Eastern European.

Mulling over Abdul's words afterward, I realized he was right. I had no grounds to judge them. I was too young to fully understand what hell they were in or on their way to, but I had an idea.

At about ten o'clock that night the place emptied out, with the men in the front rooms leaving. A couple of hours later, and my little group of five and the other fifty people in the room with us were told we too were on our way.

Outside the building was a large cattle truck with a canvas roof. It certainly didn't look big enough for fifty-five people but we were ordered in anyway. Climbing up was hard because it was so high, but I didn't want to show I was scared so I didn't ask anyone to help me. All sad thoughts of the poor women had vanished. I was too busy worrying about my own safety.

We crammed inside. Men were squashed into every free space, guards using fists and sticks to make sure no room was wasted. The four of us clung to each other so we didn't get separated. I balled down into a shape as tiny as I could make between Mehran's knees among a forest of legs and fetid feet.

I only knew Istanbul was the destination, but how far that was I had no idea.

AFTER FIVE SWEATY, PAINFUL AND CRAMPED HOURS, THE truck pulled over. A heavy-set Kurd in his mid-twenties got out of the passenger's side of the cab, adjusting his sunglasses and trucker's cap. He opened the canvas. "*Yallah.*"

Men started jumping off, and we were pushed toward some nearby trees, where some guides were waiting. The chief guide carried a large staff, just like a shepherd. "Follow me. Walk fast." He started walking uphill. And, just like sheep, we followed him.

I wanted to be sure my friends and I stayed close. It was around 3 A.M., and still dark. I took it upon myself to be the organizer: "Baryalai, I'll hold your hand. Abdul, you hold Mehran's. And let's stay to the front."

I looked back to see the truck driving off into the distance.

We didn't know why we were walking or what was coming next. The mental anxiety of not knowing where I was going or when the walking would end made every step harder: it's easier to walk for miles on end if you know where you are heading.

The terrain was a bit like the border crossing from Iran into Turkey—very rocky and steep. I had the distinct feeling that we could be ambushed and arrested—or even worse—at any moment.

We stopped to rest and drink from a little stream. It was so brief that those following up at the rear didn't have time to sit down; by the time they caught up, we'd already been ordered to get up and keep moving. My instinct to stay close to the guide at the front had been the right one.

After two hours of a forced march, we came down a little gulley into some farmland. The truck was there waiting for us, and we were ordered back into it.

I was too tired to work out how the truck knew where we were, nor how the guides had managed to bring us to the right spot.

We drove the whole of that day. It was baking hot under the canvas, and we were so thirsty and hungry. Occasionally, the driver would stop and throw a couple of bottles of water into the back. They didn't give us any food.

I was grateful for the water, but the driver didn't seem willing to stop for toilet breaks—presumably because we'd have been seen. I tried to drink as little as possible so that I didn't need to go. Some men, however, were so desperate they were peeing into the empty bottle. It smelled so bad. Eventually I had to try too, as my bladder was really hurting, but I just couldn't do it.

"Get over it, Gulwali. What's your problem? Just do it in the bottle."

"I can't."

Baryalai laughed at me. "In our company, you will become knowledgeable." Another old Pashtu saying.

I tried to smile. But I still couldn't do it in that bottle.

THIRTY HOURS LATER, AND WE WERE ON OUR THIRD TRUCK drop and hike. To the side of us, we had been able to make out that we were walking past some kind of checkpoint: we could hear dogs barking and make out queues of vehicles and soldiers. We were walking in order to skirt past them.

I understood what was happening now: the trucks drove as far as they could along the roads without meeting the police. As we got close to the police posts, they ordered us out and into the woods so that we could walk around the checkpoint. The driver would pass through the checkpoint with an empty vehicle and all his official paperwork intact. He would then drive a safe distance away, pull over, and wait for us to find him again. Then we would climb back into the truck and continue driving.

I was so thirsty and weak now that only Baryalai kept me going. He kept urging me on, taking my hand when I couldn't put another foot in front of the other. My feet were bleeding and I just wanted to lie down next to the path and sleep. But that would have been fatal.

"Come on, little man."

"I can't."

"You must."

I could not.

Our column had thinned out; I had dropped right to the back of it. Only the weakest and oldest hobbled past us now. I knew my situation was getting critical when one of the eldest men in the group wheezed by.

"*Kaka,* Uncle, help me with the child." Baryalai's voice was desperate.

"I can't."

"Help us. Look at him. He's almost done."

"I can't. If I help him, who will help me?"

"Come on, Gulwali. Get up. We've got to go. Up. Move. Move. Move."

"I'm sorry," gasped the old man, shuffling off.

Baryalai exhaled a long breath. "Here, drink." He presented the last mouthful of gray liquid from a creased Sprite bottle. "Come on, boy. You must. What would your mother say if she saw you being so weak?"

That was the right thing to say to me. I started to gather what few mental resources I had left and I kept going. One footstep at a time. Left, right, left, right—I just concentrated on the rough ground immediately beneath my feet. Left, right, left, right.

On and on and on.

After walking for four hours, we came to a place where the truck was waiting for us again. This time we drove for three hours before getting out and walking again.

I don't know how I kept going.

I was so grateful to Baryalai for looking out for me the way he did—I couldn't have made it otherwise.

In total, we did the drop and walk four times. On the penultimate drop, we sat and waited for so long for the truck to return that the sky went from late afternoon gold to the stars punctuating the sky. I stared up at the beauty of the twinkling pinpoints of light, wondering if the truck would ever return and if that night would be the night I died of cold.

The next morning, our driver dropped us by the side of a fast-flowing river and told us to drink and relieve ourselves. No sooner had we got out when two more trucks roared up in a cloud of dust. Five men climbed out and started shouting in Kurdish.

For a minute I thought their anger was directed at us, but then our driver and his mate pulled knives and ran at them. A fight broke out. We just stood and stared as those seven men punched and kicked each other. They were like wild animals—ripping off their shirts to reveal huge, hairy chests and pulling off studded leather belts, which they waved over their heads like lassos. There was a lot of shouting, screaming of insults, and blades glinting in the air. I was certain someone would get stabbed.

It appeared to be over when our driver screamed at us all to get back in the truck. I'd never been so happy to oblige.

This was a turf war. It turned out the smuggling gangs would occasionally hijack each other's groups, especially when they wanted to settle a score.

The driver got back in with a second man, who told us to be quiet and stay calm. This man said that Malik and he had a disagreement and so he was taking us as his revenge. He told us to sit tight until he was ready to order our driver where to go.

We sat there inside the truck, parked by the river, for maybe four or five hours. Waiting, expecting the worst at any second. We didn't know what to think. We thought that maybe because we'd complained to Malik about the chicken coop he'd set the whole thing up. Or maybe the second man was lying? Perhaps Malik had sold us to him? Perhaps the driver was in on it?

We all huddled in the back whispering frantically to each other.

"Should we kill them?"

"Let's run for it, while we still can."

"Just wait. Don't do anything rash."

I didn't know which way to turn. I could only hold Baryalai's hand for comfort.

In the end, after what seemed like forever, the second man got out and we continued on our way as if nothing had happened. We didn't know how the dispute had been resolved.

We were still very afraid and shaken when, a couple of hours later, the driver pulled into a remote set of farm buildings. We were told to bed down in a dirty cowshed where huddled groups of other refugees already lay. There must have been over three hundred people already there, plus our truckload. I had absolutely no idea where we were.

There were no mattresses, just straw on the floor. We were human cattle. It was a beautiful, bright autumn day but inside the sheds it was dark and dank, smelling of cow dung. We didn't understand why we were there or for how long we'd have to stay.

It had echoes of the chicken coop all over again.

MY LIFE COULDN'T HAVE FELT ANY MORE OUT OF CONTROL. Just over a month ago, I had been an ordinary schoolboy interested in my books, looking after my family, collecting firewood, teasing my aunts, and sleeping in my grandparents' bed. Those happy days already felt like a lifetime ago.

But at least my original companions, Mehran, Abdul, and Baryalai, were still with me. We'd made friends with Shah and Faizal but lost them so quickly. I was beginning to learn that people came and went, and there was no way of controlling that. And so, while I was determined to stay close to my friends for my own safety, I was also steeling myself for the inevitable loss. I didn't see the point of trying to make new friends. I didn't have the energy.

In the afternoon of the next day, one of the drivers made an announcement in the cowshed over a bullhorn: "Stay ready and alert. Your trucks are coming back this evening."

Sure enough, that evening we continued onward, driving past the outskirts of a city someone told me was the Turkish capital, Ankara. Finally, after three very long days and nights, we reached Istanbul—the city that bridges east and west.

CHAPTER 10

STRUGGLING TO BALANCE ON TOP OF A FLAKING STEEL-framed school chair, I saw the glory of Istanbul—the city famed for being a melting pot of ancient and modern—lying like a patchwork blanket before me.

I was hoping to catch a glimpse of the Blue Mosque—one of the city's iconic sights, and a place Black Wolf had told me about. His eyes had misted over when he'd described it as one of the most beautiful things he'd ever seen. But I had no idea if we were even anywhere near the mosque. All I had been able to see of Turkey's largest city for the past two days was a narrow segment of it, through a small, greasy window, a segment that was visible only when I stood on the chair. Mostly I could see just satellite dishes, television aerials, air-conditioning vents, and clothes strung across crudely erected rooftop clotheslines. I could just about make out a few minarets, but they seemed too small.

"Get off there, kid. I'm not going to tell you again."

It was the Turkish owner of the apartment we were staying in.

I hurriedly jumped down.

The night we had arrived in Istanbul, we had been taken to what looked like a shantytown. As we had disembarked from the truck, we had been greeted with the question: "People of Malik?"

I supposed that meant we were under Malik's ultimate control now that we were in Turkey.

The area was a clutter of temporary shelters made with metal poles, tin roofs, and tarpaulins. It was situated near quite a busy road and wasn't hidden by fences, so was in clear view to any passing motorists. I was amazed that undocumented refugees would be left to wait in so obvious a place—but at least there we were able to pray, go to the toilet, and walk around the space quite freely.

We waited in the shantytown for a whole day—it seems the business of smuggling takes time to organize. Malik apparently needed confirmation we had arrived in Istanbul before he paid the drivers; then the agent in Istanbul needed to get the okay that it was his turn to move us. Abdul, who was good at using his quietly calm ways to obtain information from people, had managed to glean this from the driver who had brought us here.

Once all was in order, the fifty-five people who had been in the truck were split up into smaller groups and moved off to various safe houses. Luckily, I was grouped with the three men who had been my friends since Mashhad.

The apartment Baryalai, Mehran, Abdul, and I stayed in was an anonymous-looking, two-bedroom flat in a medium-size block in a quiet suburb. We four shared one room, and another group of five shared the other. We stayed there for two weeks, not going out and getting very bored and frustrated until, without warning, we four were moved on again, to a not-dissimilar apartment a couple of blocks away.

This time it seemed we'd been passed on to three Afghan business partners who lived in Istanbul. They didn't give us much information about our situation. "We're working on a plan. You'll be in Greece soon—that's all you need to worry about."

I was even more confused when they gave me $150. "Qubat sent it, for your expenses."

To date, Black Wolf had been the only agent who had ever bothered to take the time to try to explain things—I often fantasized about how nice it would have been to stay at his farm. I could have helped him with the dodgy oil business and made myself useful. Even if I'd had to muck out his horses, it would have been preferable to the situation in which I now found myself.

By this time I was in a perpetual state of confusion that made relaxing properly, even for one night, impossible. We just never knew when we'd be told to be on the move again. I was constantly exhausted, living on adrenaline and fear.

For the first few days, we were locked inside that apartment on our own, but the three Afghans had left a few basic groceries—rice, salt, oil, tinned chickpeas—which we cooked for ourselves on a little electric stove in the kitchen. I had never so much as boiled an egg before—male children weren't expected to know how to cook in my family. Thankfully Baryalai, who had lived for several years in a refugee camp, was a master at creating a tasty meal out of nothing.

The trio of Afghan agents popped in every couple of days to check on us:

"Are you men staying quiet? Do not alert the neighbors."

On their third visit, one of them took me outside with him to get the shopping.

I was delighted to have a taste of freedom, even if only for an hour. Istanbul was enthralling, more diverse and exciting than Tehran, which now felt sterile by contrast: the vibrant noise of the city; the smell of grilling kebabs; long rows of high piles of spices and fresh herbs and vegetables outside the shops; large, smart modern blocks situated next to preserved historical buildings; and thumping pop music booming from cars. Music had been banned in Afghanistan under the Taliban as unIslamic; after they fell, pop music—mostly Indian or American—had exploded in popularity in Afghanistan. But this Turkish music was different: it was a real mixture of Arabic and Western sounds.

I was able to change a small amount of the dollars my mother had given me into Turkish lira, and I used a little to buy some treats: a new T-shirt for me, and a box of delicious dates and a bag of Turkish sweets to share with the others. I was so disappointed when the shopping trip ended and I had to go back to the house.

Climbing the stairs, I could feel the walls closing in on me once more.

THE FOLLOWING DAY, WE WERE ALL DOZING WHEN THE mobile phone the agents had left us for emergencies started shrilling.

Baryalai picked it up.

I could hear yelling down the line. "The police are coming for you. You all have to get out of there. Now."

We were given an address, told to split into pairs, and walk calmly down the street to it. We shoved our meagre belongings into our bags and rushed to leave the flat. I walked with Abdul, and Mehran was with Baryalai.

The new address wasn't far away. I wondered how many other houses and apartments across Istanbul were also hiding the displaced and desperate.

For the next month we were moved around a lot. Everywhere we stayed there was always a succession of bitter, angry, tearful refugees who found themselves sliding downward on this game of Chutes and Ladders, on journeys that had stalled or gone awry, or were going backward as they were forced to return to their homelands. They couldn't wait to share their tales of woe and warnings of what was to come for the fresher-faced hopefuls.

One man we met in the basement told us he'd recently escaped from a kidnap situation in Kurdish Iran—where we'd just come from. He had been forced to write home to ask for more money, but no one had replied and the smuggler wouldn't release him. It had taken three months for him to sort it out so that he could reach Istanbul and continue the journey. Others had been tricked into paying whatever cash they had left to take fake routes, or pay for transport that never arrived, leaving them stranded. Others had faced arrest and deportation. Some had been immersed in this nightmare for so long they had almost forgotten who they were or where it was they were supposed to be going.

I listened to truly terrible tales of beatings, blackmail, women and children suffocating in trucks, or men being shot at point-blank range when they argued with their handlers. In some ways, I suppose it was helpful to be warned, but it only made me more scared: knowing that I might be tricked or physically hurt didn't mean that I could prevent it. I was at the mercy of the drivers, guesthouse owners, and their bosses—the agents of the actual smugglers. And there was nothing I could do about it.

Our current three agents continued to tell us they were working on a plan to get us to Greece. Every time they checked in on us we heard the same promise: "Tomorrow you will go to Greece."

But tomorrow was taking an awfully long time to come.

GREECE WAS A MYTHICAL, MAGICAL COUNTRY I HAD READ about in school. I knew it as an ancient civilization and the home of Alexander the Great, a man respected in Afghan history as a great warrior after marching through central Afghanistan in 330 BCE. He famously wrote a letter to his mother about the bravery of the Afghan warriors he fought: "You have brought only one son into the world, but every man in this land could be called an Alexander." His blond-haired, blue-eyed descendants are still very visible in the areas where he and his men made camp.

As a schoolboy, the idea of visiting such a historic place as Greece would have been the dream of a lifetime.

But not like this.

Other refugees told us that from Istanbul the smugglers might take us one of three ways to Greece, and then on to mainland Europe.

The easiest was across the Aegean Sea by boat. I heard that if you were lucky it would only take three hours, yet the journey was fraught with terrible dangers on overcrowded and unseaworthy vessels. As a child from a landlocked country who had no idea how to swim, this was beyond the wilds of my imagination. But if that sounded bad to me, the second route was worse—one hollow-eyed refugee described it to me as "the pathway to hell itself." This route had to be done on foot, marching over dangerous mountains, hiding near borders, and crossing three countries, from Turkey into Bulgaria, then on to Macedonia, and finally into Greece. The third path was also an overland route, crossing a narrow, 12.5-kilometer strip of heavily militarized land that forms the direct border between Turkey and Greece.

Everywhere we stayed, other refugees filled us with fear about the overland routes.

"No roads, no tracks, only rocks and mountains."

"No way to get food when it runs out."

"Bodies were littered all along the pathway. No one will ever find them."

"It's a wasteland of death."

I heard so many terrifying descriptions from those who had tried to make it, but had either been turned back by Bulgarian border guards or forced to retreat by the vicious weather, that I must have crossed both overland routes hundreds of times in my head. As scared as I was of the thought of a sea crossing, these other two routes filled me with absolute dread. I was told that both these routes could take weeks or even months to navigate, and that there were many rivers with terrible currents. Some you could only cross with a small speedboat, but the boats the smugglers provided were very old and they often capsized—and that meant certain death, dashed on the rocks by the fast-flowing waters. If there was no boat, you had to try to cross on foot. I was told of smugglers who had forced screaming women and children into the water at gunpoint, only to watch them drown. Moreover, portions of both routes were said to have deadly landmines or anti-vehicle mines. I knew all about landmines; they were one of the tragic legacies of the Afghan civil war: some ten to fifteen million mines were said to still litter the country in the year I was born. One of my closest friends at school only had one leg—the other had been blown off after he had stepped on a mine during a football game. And, according to pretty much everyone who had been that way, Bulgarian border guards were notorious for shooting at refugees. If you were lucky enough not to be shot, they might beat you before stealing your possessions and forcing you to walk back to Turkey.

The worst story of all was told to me by a Pakistani man. He and his brother had attempted the third route in the middle of winter. They had become separated from their group and wandered through the snow until they were rescued by Greek soldiers. His brother had had such bad frostbite he had been taken to a hospital by the soldiers, where his fingers were amputated. After he recovered, both brothers were deported back to Pakistan. And now the one telling me the story was here again, attempting the journey for the second time. "When we went home, my brother wasn't sad for the loss of his hand. He was sad because he became useless, unable to support our family. I have to try again because if I don't make it, who will feed us?"

Hearing this story caused my brother Hazrat to haunt my dreams more than ever. Where was he now? Had he been tricked into taking one of the overland routes? Was my brother's broken body lying on the bottom of a riverbed somewhere? Had he been shot?

Was he lying out there bleeding, calling for me?

CHAPTER 11

I WAS ALMOST BEGINNING TO WANT TO BE CAUGHT: BEING deported surely had to be better than getting killed. But the thought that Hazrat was out there somewhere, lost and alone, stopped me. Hazrat would be looking for me too, I was certain of that. Besides, if I went home without him, I knew my mother would never forgive me.

The only solace in my existence was that by now the three Istanbul-based Afghan agents trusted our little group and gave us more freedom, so we were able to go outside for a few hours a day. We walked through parks or drank Turkish coffee in pavement cafes—anything to help pass the time. On one of our walks we discovered an Afghan-run DVD rental and sale store, something that had Baryalai and Mehran practically whooping with joy. Films had also been banned under the Taliban but, once they had gone, bootleg movies had flooded the bazaar in our town. I hadn't been able to see them, however, because my family didn't have a TV, and my father didn't approve of such things anyway.

After chatting to the friendly shopkeeper at the store, Baryalai used some of his precious cash to purchase three copies of Bollywood movies. He was as excited as a little kid. "I saw this one three times in Peshawar. It's so good. Trust me, little man, you are going to love it."

When we got back to the *musafir khanna,* he fiddled with the Chinese-made TV and DVD player that stood in a corner. It had a label on it which read: "Sonysonic." Baryalai told me that was the maker. "Don't you get it? It's brilliant. The Chinese are so enterprising." He laughed, but I couldn't understand why he found it so funny.

The films were okay. They were in Hindi with English subtitles, so I couldn't understand what was being said. There was a lot of singing and dancing and a big fight scene. I couldn't really work out what was going on except that it was some kind of love story. I very nearly nodded off. When the closing credits rolled, I looked over at my friend. He had tears running down his face.

The next film was a historical action movie. Even though I still couldn't understand the words, I quite liked the fight scenes, and I much preferred this movie—so much so that by the time it ended, I was so utterly engrossed in it I had completely forgotten it was make-believe. The final epic battle had seen so many brave warriors fall that I turned to Mehran and said, "Is everyone in India dead now?"

He bopped me over the head.

Those few moments of light relief were rare. By now all four of us were completely fed up with the constant moving and changing of locations, and the anxiety and uncertainty it created. We complained to one of the Afghan trio. I begged them to contact Qubat and see how my family was.

"We'll try, but he'll probably be too busy."

"But I need to tell them how I am."

They wouldn't make the call for me but they did reassure me my family would know I was safe and where I was. They told me that every time I crossed into a new country my family would be informed by Qubat or one of his representatives. This was because his next installment of payment would be due. They explained the money my family had paid was held by a mutually trusted third party, a kind of smuggling layaway plan. Each time I crossed a new border, the third party handed over a little more to Qubat.

That didn't make me feel better, and I suspected they could have been lying to me. In fact, we complained so much that the trio said they were sick of us and were passing us on to a different Istanbul agent. We had little choice but to do as they said. All we knew—all they told us—was that we were expected to meet the new agent in a certain cafe.

The new agent was another Afghan: Zamir. He was young, and smartly dressed in Western-style clothes. He wore a casual, open-necked shirt and gel in his fashionable haircut. Leather bracelets were wrapped around his wrists.

"I'm taking you to a really good place. It's the best *musafir khanna* in all of Istanbul."

I was coming to realize that the agents were the salesmen of the smuggling world. They have to sell their fat lies and thin hope to convince you to keep going—it's in their interest, because if they don't get you to the next location they don't get paid. They will say anything to persuade you. But it's the people below them—the employees, the drivers, the farmers with the cowsheds where the agents hid us—these were the ones that were the most brutal. They had already been paid by the agents, and so had nothing to lose. Often they subcontracted their work to family members or friends. That's when it got really messy and you had no idea who was whom.

Having to work out all of that was making me grow up very fast.

THERE ARE EXAGGERATIONS AND THEN THERE ARE COMPLETE lies. As soon as Zamir delivered us to the new address, I knew it was a bad place. Our new hideout was guarded day and night by three nasty-looking Turks, who glowered at us menacingly. We didn't see weapons but I am sure they had them, because one of them kept fingering the bulge in his pocket.

It was another basement, already packed with twenty to thirty Afghans, who didn't waste a second in letting us know what a mistake we'd made by coming there.

"This man is a liar. Everything he says is pure bullshit."

"Months. I've been here for months."

"We are less than human."

Every time someone tried to speak, the thugs ordered silence.

I was so depressed to be there. The basement had tiny, locked windows. There was no fan and it was brutally hot inside. So many times I thought I was going to faint. No one was allowed outside and the only food was takeout, brought in daily by the thugs. And we had to pay for that with our own money. The others had been borrowing money from each other just to live.

After one particularly noisy and restless night, one guy got up and announced he'd had enough. "Fuck this, I'm leaving."

The guards punched him to the floor. "Think we'll let you go and call the police?"

My exit came when Zamir arrived one day and read out a list of ten names. He called my name, but not my three friends.

"Don't go, Gulwali," they all urged. "Stay with us so we can look after you."

I didn't want to go, but I didn't want to get stuck there for months either. I swallowed hard and summoned my courage. Looking Zamir square in the eye, I said, "If they don't go, then I don't go."

"Fine, it's up to you," he replied, calling my bluff. "They're not going. Stay if you want. But let me tell you, this is a guaranteed trip."

That made me stop and think. This was *my* journey. I *had* to continue. And I also wanted to prove something to myself: I had heard people whispering behind my back, saying how my friends had to look after me and what a little boy I was. It made me defensive and angry. I was determined to show I could manage on my own—to prove I was grown up.

I went.

Zamir's men shoved us roughly into the back of a van and drove us to a train station. I had a very bad feeling that this was not the way to Greece—not by boat anyway. The train station was closed for the night but the railway security guards opened some gates at the back to allow our van in. They greeted our agents as if they were old friends and ushered us toward the rear of the station, where an empty train sat in the sidings.

We hadn't had food or water all day and I was wobbly with hunger.

"You. The small one. Come here." A guard motioned me toward the train and into the driver's cab. I clambered inside and took a seat.

"No, not there. Here." He pointed to the ceiling.

"What? I don't understand."

He laughed long and hard, clutching his fat belly with a scaly hand. Then he stood on the seat and unscrewed a large acrylic panel that covered the ceiling light. "Up here."

How was I was supposed to fit in *there*?

Warily, I climbed onto the seat. I tried to haul myself up as the guard gave me a vigorous shove up my backside. The cavity was pitch black and chokingly full of dust. My eyes could just about make out a tiny hollow next to the wiring of the light box. Surely not?

"Get in. Hurry up."

With that he gave me another hard shove. By wiggling, I just about got my torso in. "I can't get my legs in. It's too small."

Another belly laugh. "Get in."

Somehow I managed to twist at the waist, contorting my legs in behind me. My eyes and nose filled with choking black dust and grime. It felt like a coffin. My coffin.

I cried out in pain and fear as the guard stood on the chair again and began to screw the ceiling panel firmly back in place. "Please. No. Let me out. Please."

"Shut up, boy." He replaced the last screw, locking me firmly into the claustrophobic, filthy blackness.

"BREATHE, GULWALI. KEEP BREATHING." TALKING TO MYSELF made me feel calmer. "Stay alive. You can do it. Breathe."

I realized I was heading to Bulgaria.

The train had begun to move. I didn't know if any passengers had come on board because I couldn't hear anything save for my own panicked breaths.

I think I managed to breathe myself into some kind of trance because I don't really know how long I was up there. I think it was a long time. The next thing I knew, the conductor was unscrewing the panel, revealing a dark landscape flashing past. "Get down. Come on. Down."

I tried to move my legs but they were stiff with cramp. "I can't."

"For God's sake, boy. Hurry." He grabbed at my legs, yanking me back through the tiny space.

I fell out, bashing my torso on the sharp edge of the seat before hitting the floor. My body went into shock. I was gulping for air, hyperventilating.

The train was still moving. But where were the others?

"Hurry. Jump. Jump when I say."

"Jump? What—? I don't . . ."

Grabbing my shoulder, he yanked me up and toward the door of the driver's cab. He flung the door open to reveal ground moving swiftly beneath us. Rocks, grass, and fields swam by my blurred vision in the twilight.

If I jumped I'd surely die.

I looked to my right and saw the others, all standing in the doorways of the next carriage. I could read the terror on their faces.

"I can't. Please, no. Don't make me," I pleaded to the conductor.

"There's a checkpoint coming up, you stupid little fool. Get off my train."

As the train rounded a bend it slowed slightly. He tried to push me out. I fought back, trying desperately to cling on to both him and the train.

"Gulwali, look out. Jump."

One of the others was pointing up the tracks ahead of us. The landscape was changing. Grass and trees gave way to large, jagged rocks.

Someone shouted, "In the name of Allah, jump now. Everybody jump now."

I don't know if I jumped of my own accord, or if the conductor pushed me. All I remember is hurtling through the air and seeing the earth moving underneath me before crashing down in an agonized heap.

I think I must have blacked out. When I came to, my head hurt so much I could barely lift it. Everyone else was screaming. It was still dark, and my eyes struggled to make out the shapes.

"My leg. Allah. Help me. My leg."

Slowly I eased myself up. The train was hurtling into the distance. Scattered along the track were nine bodies in various states of brokenness.

"Gulwali, help me. My leg."

I staggered over to Zia, an Iraqi. His leg was twisted at a sickening angle.

I retched before collapsing and starting to cry. My head hurt so much I thought I was dying.

Two of the others appeared by our side. "Where is everyone else?"

People were still screaming uncontrollably. Some of them looked really badly injured.

After that it all became a blur. The police arrived with an ambulance at some point. The badly injured were taken away and I never saw any of them again.

I was semi-conscious as I was bundled into a police car.

CHAPTER 12

When I awoke, I was alone in a Bulgarian police cell. Someone had placed me on a rickety iron bed with just a thin mattress and no bedding. Through the bars of my cell I could see others from the train lying in similar beds, in other cells.

Istanbul had been so hot and stuffy; here, the air was freezing, but stunk of rotting flesh. And it wasn't just refugees in there; some of the other prisoners looked local and really scary. I don't know if they were mad or drunk, but a few of them shouted out all night long.

My head was hurting so much I could barely move. If I tried to stand I felt dizzy and had to lay back down. I tried using hand gestures and a mixture of different languages to appeal for help. "Please. I need a doctor. Help me. I'm begging you."

The guard who brought me my meager rations of food refused to even look at me.

After a couple of days, they told us we were going back to Turkey. I knew what to expect because I'd heard it all already from the stories: the Bulgarians sent people back over the border without the knowledge of the Turkish authorities.

From Sofia, we were put on a police bus and driven for several hours to a forest. Half a dozen guards pointed to the trees—we were expected to walk back to Turkey.

I think it was probably around January or February by then, and there was heavy snow on the ground.

We hesitated. They pointed their rifles at us.

It was pretty clear we had no choice but to move.

We walked into the trees, not knowing what direction to take. It was so very cold. My boots—"my best friends" as I jokingly called them—were holding up well, but my thin clothes were filthy and ripped after jumping from the train and our stay in the cells.

Of our party, there were three who had leaped from the train with me, some of the least injured. I didn't know about the others from the train—most likely they were at a hospital, but I couldn't say for sure. The rest of the shambling column was made up of other refugees from the prison.

How I missed my three friends. I missed their company, and their help and support. Maybe I wasn't as grown up as I had tried to convince myself I was.

We walked through the snow for several hours. My toes went numb, along with my face and fingers. I wondered if my fate would be that of the Pakistani man who lost his fingers. Would I go home to my mother with some terrible injury, rendering me a burden for the rest of my life?

Eventually, the trees began to thin out and we reached a small village. A flag above one of the rooftops told us we were back in Turkey. Although I was going backward not forward, I didn't mind; I wanted to cry with relief at having escaped Bulgaria: the train had clearly been the first step of that notorious overland route to Greece via Bulgaria and Macedonia that I'd been warned about. The train had almost killed me; I'm certain that I would not have survived the rest of the journey.

At the village, the four of us from the train separated from the rest and got a taxi to a larger town. From there, we used my Turkish lira—the money I had exchanged on the shopping trip with the Afghan agents—to buy bus tickets. The other two only had dollars, but they promised to pay me back once we reached Istanbul.

Before we got on the bus, I used a pay phone to call Zamir. I had his number written in my pocket on a piece of paper he'd given me. He had said he'd send someone to the station to meet us.

"But why aren't you in Bulgaria?"

This made me so angry. "Why aren't I in Bulgaria? Because one of your friends threw me off the train. Then we got put in prison. Then we marched through the snow like the Soviet army retreating from Afghanistan. How could you send me on my own like that? That was not good of you."

"Calm down, Gulwali. You're all right now, aren't you? You did agree to go. Don't you remember?"

"I agreed to go to Greece. Now I am in Turkey."

"Well. This was the way to Greece. So it didn't work. Not a big deal. We'll try it again."

"I am not going that way again. People were hurt."

"Really? I must say that is very unusual. Let's not discuss it now. We'll talk about it when you return."

He sent one of his thugs to pick us up, and we were taken back to the same place as before. Baryalai, Mehran, and Abdul were still there. I was so relieved to see them.

THE DAY AFTER MY RETURN FROM BULGARIA, ZAMIR PASSED us on to another agent, an associate of his, a Turkish Kurd.

He came to pick the four of us up in a taxi. I was just grateful I wasn't being separated from my friends again.

This new agent didn't bother to tell us his name. He was about the same age as Zamir and similarly dressed, in a tight open-necked shirt, denim jeans, and pointy leather shoes. He reeked of cheap aftershave.

In the taxi, he spoke to us in what was by now becoming a typical mixture of Farsi and Kurdish: "We need to wait until night. Then I will take you to a place where you can get the boat to Greece. We will wait at a place I know."

Given what had happened to me in Bulgaria, I did not trust a word he said. And I knew for sure we'd been lied to when he directed the taxi to a shop in the middle of town, opposite a petrol station. This exact location had been described in minute detail to us by other refugees. It was the smugglers' holding center everyone dreaded the most, one of the places they kept you before sending you overland on the second route—the journey that took you over the heavily militarized Turkish–Greek border.

We were led up a metal staircase at the back of the shop to a scruffy apartment. There were already forty or so refugees there, who looked to be a mix of nationalities. Everyone was sitting on the floor in silence, their heads down.

We tried to get away before we got trapped: as we neared the door and saw the scene inside, those at the front turned to go back down the stairs—but the agent was standing behind them, blocking their way. Abdul had grabbed my sleeve to try to pull me back when a guard by the doorway yanked me by my arm inside.

We were pretty much forced inside and told to take our places on the floor with the other people. The Kurdish agent spoke briefly to the guards then left us. We never saw him again.

There was no food or water, and it was freezing cold despite the number of densely packed bodies. The atmosphere was awful, depression and sadness hanging in the air like a black cloud. I think every man in the room had an idea of the fate that was about to befall us.

I sat as close to my three friends as I could; no way was I being separated from them now. We sat in silence, lost in our thoughts, for a couple of hours. I tried to perk myself up, telling myself I might survive this, and that I had to try, at least. But it didn't really work: my thoughts kept taking me back home—I thought of my little siblings and how they were growing up without me; I thought of my second brother, Noor, so close in age to Hazrat and me, but who had been left behind to become the man of the house, aged just eleven. How had he reacted when he was finally told that Hazrat and I weren't coming back from our trip to Waziristan? How betrayed had he felt when he was told he'd never see his older brothers again?

"Psst."

I looked up. Abdul was trying to get my attention.

"Shut up." The guard glanced at him in warning.

Abdul ignored him and continued to whisper: "What are we doing, sitting here like dead men? Let's just go. They can't force us to stay here."

He was right. My eyes met Baryalai's questioningly. He nodded.

I stood up first. I almost wanted to laugh. Abdul was right. Were we lambs to the slaughter? No. We were human beings, free to choose our own destiny. We would stroll out of there and find a way to manage on our own.

The cupped palm collided with my ear, sending a shock wave of pain through my skull. I fell onto a patch of dirty carpet and rolled into a ball, seeing the guard slump back down heavily onto his chair through watering eyes.

"Leave the boy alone," said Baryalai.

I lifted my head just as the guard's leg flashed like a cobra, sinking its teeth in the small of my friend's back. He let out a yelp of pain and writhed on the floor, clutching a hand to his spine.

"Sit down."

Bang. Bang. Bang.

The room was suddenly flooded with panic.

Bang. Bang. Bang.

"Someone's at the door."

"Shut up—they'll hear you."

A hollow voice boomed in Turkish. We were trying to stay quiet so no one translated for me, but I could read the others' reactions and knew enough words of Turkish by now to work it out.

"This is the police. We know you are in there. Open the door now."

The guard replied in Kurdish: "I can't. I, er, I lost my key." He turned wide-eyed to his colleagues, shrugging his shoulders. They nodded at him, gesturing at him to carry on. "My friend locked it when he went out, forgetting that I am still in here. He is a bit stupid like that sometimes. He will be back later tonight—you really shouldn't trouble yourselves to wait so long."

There was no answer. Instead, an ax head split the door in two, and then a dozen men stormed into the room, shouting: "Police. Don't move. Lie on the floor."

They had guns and wore bulletproof vests, but weren't in uniform, which only compounded the sense of panic.

"Kidnappers."

Some people tried to escape through the back, or get to the window. The men screamed at them and pointed their guns at them, telling them not to move.

Uniformed officers then flooded into the flat, adding to the chaos.

Men flung themselves out of the way in terror, but there was nowhere to move to. I half expected the police to start shooting at any second.

I'd just been in a Bulgarian prison, and I didn't want to be jailed again but, as I placed my hands on my head as they demanded, I felt a surge of pure relief and gratitude. I definitely hadn't wanted to go on the overland route.

We sat on the floor, arms on our heads, for an hour or so. Eventually, we were loaded into the back of a police van and driven to a large police station and prison complex somewhere near the center of Istanbul.

We were asked our name, age, and country of origin, then taken up two flights of stairs, where we were seated in a large hallway. There were no chairs so we sat on the floor, tightly packed together. A couple of officers stood watch. With the lack of space, the way we were sitting in silence, and the guards, it didn't feel much different from the place we'd just been rescued from.

Selling tailor supplies at my stall in the bazaar, age 8.

OPPOSITE: As an apprentice tailor, age 10, with my brother Nasir. I am wearing clothes I made myself.

ABOVE: A familiar landscape: Laghman Province, Afghanistan. *Photograph by Naveed Yousafzai.*

With my brother Noor (left) and my uncle and cousins, 2002.

Afghan passport photo, 2008.
Kent Social Services was disputing
my age at the time.

With my foster father, Sean. This man has been one of my greatest inspirations.

Me at school in England.
At this time I was so depressed
I wanted to kill myself.

Carrying the Olympic torch for my adopted homeland was the most amazing experience I had ever had. Burnley, June 2012.

SIX HOURS LATER I WAS QUIVERING, ALONE, AND AFRAID, IN an interrogation room. Somewhere over my head a striplight made a little *plink-plink-pink* sound, a bit like my grandfather tuning his *rabab,* an Afghan lute. I could see the stuttering bulb in the interrogation room's mirrored walls, and I jumped each time I caught my own reflection looking at me—as if the red-eyed stranger staring back at me might take offence.

A young, uniformed officer entered the room. "So," he said, scraping his chair on the floor as he pulled it out, and slapping a notebook down. He spoke in Turkish, but the basic questions I could understand.

"What is your name?"

"Gulwali."

"How old are you?"

"Twelve and a half."

"Where are you from?"

"Mauritania."

He sighed. Then he turned to a second man, who was not in uniform, and shook his head.

The second man spoke to me in Farsi. "You are our seventeenth interview this morning. Are you going to waste our time with lies, too? Or are you going to be a good boy and tell me the truth?"

I understood the translator perfectly well, but I had to stick to my story. Other refugees had warned me that if I ever got arrested in Turkey, I should say I was from Mauritania. I didn't know why, or even where the place was, but word was that they might let you stay in Turkey if that's what you said.

I pretended I didn't understand him and started waving my arms around, as if I was doing sign language.

Now it was the translator's turn to shake his head. "I know you understand me. Where are you from?"

"Mauritania."

The notebook slapped me in the face. Suddenly, I could feel tears coming. I was a young boy searching for safety—my only crime was to be traveling without the right documents. He let out a little laugh. "Mauritania, you say? So, let's continue this conversation in French. *Je parle assez en avoir assez pour vous comprendre.*"

Why was he saying that? I didn't speak a single word of French.

I just continued to wave my arms. I hoped, if I was lucky, he might think I was mentally ill.

He paused, gave a sneer at my obvious incomprehension, and called out sharply to the mirrored wall.

The door opened. I watched in the mirrors as two more uniformed officers, tall men with hard faces, walked in and stood behind me.

My mouth went dry.

"Take him outside."

The pair walked me back up the stairs to where the others were sitting. I made to sit back down.

"No. Keep walking."

They walked me up another two flights until we stood at the very top of the stairwell. There, they spun me around, and while I was still fighting for balance, shoved me backward. "Look down."

Three strong hands gripped my skull; the fourth grabbed a fistful of my hair at my temples and twisted hard.

It felt as if they were peeling the skin from my head and I screamed—more in shock and fear than pain.

Then they marched me back down to the interrogation room to face the first two officers. But this time the second pair stayed, standing over me.

I told them the truth. "I'm from Afghanistan. I am Gulwali from Afghanistan."

One of the men behind me threw me forward suddenly so that my cheekbone crunched onto the tabletop.

I looked up at the two interviewing officers, dazed and petrified. There was no holding back the tears. I mopped my eyes and nose with the back of my grubby sleeve.

"Tell me about the smugglers."

"What?" I said. The question didn't seem to make sense.

"Where did you stay?"

"I don't know. I don't know, I promise."

The hands locked themselves to my head again, twisting my hair so hard it came out.

"Please. Stop it. It hurts."

"Who were the men? Who did you pay?"

"I don't know. Please. I don't know."

They released their grip and my head fell forward again. I spluttered tears and snot all over the table.

"Get him out of here. Let's see what the next idiot has to say."

I was taken back to the waiting area, where several of the men had swelling to their faces or bleeding lips.

Every so often, the officers would drop a sweaty man off and pick a fresh candidate to take with them.

Mehran was sitting next to me. He hadn't been interviewed yet. "Did you say Mauritania?" he whispered.

"Yes, but it doesn't work. Don't even say it."

"Oh, so you don't want me to be free?"

"Go ahead then, say it. Tell them that. If you like pain."

It was several hours later, almost morning, by the time all the interviews had finished. We had been sitting there without food, water, or sleep all night.

A young man in a shiny blue suit appeared on the stairwell, flanked by the Farsi-speaking officer. "I am from the Afghan embassy." Hopeful eyes turned to look at him. "You are to be taken to prison, where you will be held until the Turkish authorities are ready to deport you."

"Please," someone said, "can't you help us?"

"You are here illegally, without documentation," said the diplomat, adjusting his dark-gray pencil tie. "What do you expect the embassy to do?"

"We expect you to do your job. We are Afghans, your citizens. You are sworn to serve the nation," Baryalai spoke up suddenly, his face flushing with anger.

"Don't," I hissed, mindful of the interrogation room and the hovering guards.

The diplomat swallowed hard. He looked at Baryalai. "I am sorry, my friend. But do you have an Afghan passport on you?"

Baryalai said nothing.

Satisfied his tie was straight, the diplomat spun on his leather heel and clattered off down the corridor.

The guards waited a moment, then grabbed Baryalai by the shoulders, pulling him to his feet.

He struggled for a moment. "Get your hands off me."

This didn't feel good. We had no idea where he was being taken. "We'll be thinking of you. Please be careful," I called to him. A guard looked over his shoulder at me and I dropped my eyes, rubbing my throbbing scalp.

Baryalai was our unofficial leader. Abdul, Mehran, and I stared at each other helplessly.

We were then ordered to stand up and told we would be taken to the cell block next door.

The holding cell was huge. It was packed with men of many different nationalities, who sat huddled in ethnic groups of African, Arab, Persian, or Asian descent. There were a few bunk beds around the corners, all occupied.

Many of the faces bore witness to recent questioning.

We found a small piece of empty cell near the door, the only floor space available.

"Please, can we have blankets?"

"No."

As I tried to go to sleep on the concrete floor, my head still aching from the interrogation, misery began to creep in. Was this really what my mother had sent me away for? How could this be better, or safer, than what I had left behind? Would the foreign forces in Afghanistan have treated me worse if they had been questioning me? I wanted to believe so, but the cold floor sucked any certainty from me.

IN THE MORNING, I WOKE TO FEET STUMBLING OVER ME. Food was being served right where we were lying. We had to get out of the way, fast.

I looked around the cell properly for the first time. I could see the faces of the whole world there, each nation sitting in its own area of the room.

"Get some food before it goes." Mehran handed me a little metal bowl of rice.

I fell on it, wolfing it down gratefully.

For the next few days we were told nothing. I had no idea what was happening, but some of the prisoners did seem to show genuine concern for me.

"You should be in the kids' prison."

"What are you doing here with us?"

One of the prisoners, an Iranian called Bernard, was small, dark-haired, and kind. He told us he was sometimes asked to help translate during interrogations. As such, he had information for us.

"Your friend. He has been taken to another prison."

"Baryalai?"

"Yes. The guards suspect he is an agent. It is very serious for him."

"Are you sure?" I asked.

"This is what one of the guards told me. Maybe he lies, but . . ." He paused. "I thought you would want to know."

A few days later, we were taken to a shower room. There were only a handful of them and we had to queue for ages. It was the first time I had seen a Western-style toilet—the type you sit upon. This one was filthy.

None of the shower doors had locks, so Mehran and I took turns standing guard for each other. In my culture, this was a very humiliating thing. I was so embarrassed having to do such private things surrounded by so many people. It was the first shower I'd had for two weeks, but it wasn't enjoyable. In the end it was a matter of necessity, rather than pleasure.

The water was freezing cold and there wasn't any soap, but it did feel healing. My skin tingled. I washed my bruised face and head with care, and rinsed my mouth repeatedly, running a finger over my teeth and gums.

Back in the cell I did a little mental audit. I had about $250 left. Some had gone in Turkey—a haircut, some new clothes. More clothes in Iran. I had spent money in Turkey and in Istanbul—mostly on food.

The prison was overflowing with refugees, the overwhelming majority of whom were Sudanese or Somalian. I hadn't been around very many black people in my life; in Afghanistan there aren't any, and although I'd seen a few Africans around Istanbul, I hadn't got to know any. I felt so sorry for the ones in this prison—some of them had been here for years. I was told a few were there on drug charges.

I was managing to hide my money in my underwear, and I had some in my pocket. I had also given $50 to Mehran to look after for me. That got stolen.

Mehran was upset and angry. He thought maybe I thought he was lying to me and that I thought he'd taken it. Of course I didn't. I totally trusted him.

We went to see Bernard, the kindly Iranian. He shrugged his

shoulders sadly. "These things happen in here. You have to accept your money has gone."

After the horrible beating, days of little food, and even less sleep, the disappearance of my money was too much. They were dollars my mother had given me. They had been a symbol of her hope for me, and the future she wished me to have. And now I had lost them. I couldn't stop crying.

"What's up?"

I stared into a round, black face and a pair of the deepest, kindest eyes I had seen in a long time. Seeing a black person was still very unusual for me.

"Why are you crying?"

"It's nothing," snapped Mehran. "Leave us alone."

"It doesn't look like nothing. Tell me."

The shock of him speaking a Persian language made me forget my tears. "How can you speak Farsi?"

He laughed and threw his arms wide. "I study here at this fine university. Nothing else to do. So, tell me. Why do you cry?"

"Someone took my money."

"Who?"

"I don't know." I began to sob louder. "My mother gave me that money."

"I see. How much?"

"Fifty dollars."

"I see. I'll sort it out." He gave me a concerned stare. "What's your name?"

"Gulwali."

"I am Marrion. Don't cry, little one. No more tears, okay? I will sort this out. I promise."

I stared at the floor and bit my lip. When I looked up again, I was surprised to see Marrion's tall form already striding back across the cell.

"Do you think he'll get it back?"

Mehran seemed unconvinced.

To be honest, so was I.

I lay down and fell into a depressed slumber. Proper sleep was all but impossible on that cold floor.

"GULWALI."

The voice crashed into my lovely daydream, in which it was Eid, and I was breaking fast with all my family. We were eating fat and juicy fresh dates. I was happy. We all were.

At the sound of my name, my family vanished.

My new friend Marrion stood over me. "Here," he said. "I found it for you." He tossed a crumpled ball of green-gray paper to me, and I clutched my money to my chest. "The thief says he is very sorry."

"Thank you." I nodded and smiled at him with gratitude.

"It was my pleasure. See you later. No more crying, okay boss?" He strode back to the African corner of the cell without another word.

I was so happy; suddenly, the world felt like a good place. Mehran had cheered up, we had a new friend in the horrible prison, and I had my money. This money—the exact note that my mother had given me—was my last connection to her. And now I had it back. That was all that mattered.

There was some kind of commotion going on with the guards outside in the corridor, but I ignored it. I was exhausted from all my crying so I tried to sleep again. I was just managing to nod off when I heard some Pakistani prisoners telling Mehran about something.

"No one could believe it."

"I know. When he started hitting his head on the sink I thought he was going to kill him."

One of them threw me a glance. "You missed out on all the fun, boy."

"What happened?"

"Two African guys had a massive fight in the showers."

"What? Just now?"

"Yeah. Only, it wasn't so much a fight. One of them beat the living shit out of the other. Only stopped to take something out of the other guy's pocket and then went."

"The guards did nothing. We were waiting outside and they stood there watching, like us."

I was too scared to ask.

"Gulwali? Weren't you listening?" Mehran had clearly worked it out. "The big guy really gave it to him. He didn't look good at the end."

I felt vomit rising in my throat.

Marrion.

CHAPTER 13

THE NEWS SPREAD LIKE WILDFIRE THROUGHOUT THE block.

We couldn't have known it before because we hadn't been there long enough, but it turned out that Marrion was one of the longest-serving prisoners there, and pretty much ran the place. He was in on drug-smuggling charges and no one dared cross him, not even the guards. He claimed he'd killed hundreds of people. He came from the Democratic Republic of Congo, where it was rumored he was wanted for war crimes, although some people said this was a whisper he himself had put about, so I don't know for sure if that was true. He also spoke seven languages fluently.

I was sick to my stomach at the knowledge that a man had been beaten because of me. The money burned in my pocket like acid. Was this what a man's life was worth? Fifty dollars?

Marrion was now a bad man in my eyes. He shouldn't have hit the other man. And yet, he'd got my money back for me. And he'd been so kind to me . . . I couldn't believe I'd imagined the softness and concern in his eyes.

Not only had the story of the beating spread, but also the reasons for it. It meant that every man in the cell knew I had money on me—but now that it was known I was under Marrion's protection, no one dared touch me.

I spent most of the next day on my knees, praying for the other man and asking for forgiveness.

PRISON WAS AN ANGRY AND FRUSTRATING TIME. THE BOREdom was terrible—a waiting game for something unknown. Most of the men played card games or chatted. Bernard had kindly loaned me a Quran, from which I took great comfort whilst reading it. I spent much of my time praying, walking around the cell, thinking about life, and reflecting on the purpose Allah had for me. It felt like a test of my faith.

I met men from Iraq, Iran, India, and Pakistan, and from all across Africa. Some of them, especially the Africans, told such terrible tales of poverty and hardship it made my own life sound easy by comparison.

Many were dirt-poor fishermen or farmers. They had left behind their wives and children because they couldn't make enough money to feed their families. I heard stories of toddlers dying of hunger in their parents' arms, of wives falling sick with a simple fever and dying for lack of medicine, of babies stillborn to malnourished mothers. The only chance of their family surviving was if these men made it to Europe and found enough work to send money back home.

But not all were poor like that. Many of them were very inspirational people to me: lawyers, doctors, teachers, journalists, or engineers. Often these were people who had been denied work or persecuted because of their political beliefs.

The most tragic story was that of my friend Bernard. He was a university professor from Shiraz. He told us he had helped to run an underground school, without government permission, educating Afghan refugee children. He told us he'd written a paper that was critical of the government and he'd been arrested and tortured. After being released he had fled the country, to Turkey.

I heard him crying on his bunk one day and I asked him what was wrong. Normally, if you saw someone upset you wouldn't say anything to them. Everyone kept respectful distances from emotion: we all knew we had our problems, and prying wasn't the done thing. I didn't mean to be rude by intruding, though—I just wanted to know. I'd never heard a man cry before, not in this way. He was sobbing so softy, yet as if the world was utterly without hope.

"Why do you cry, Uncle?"

"They are sending me back to Iran."

"But you will be going home and out of this jail . . . maybe that's a good thing—?"

"They say that I damage the minds of my students. They made me confess—electrocuted me—and made me sign my name, like a criminal. I escaped once. This time I won't. They will hurt me again. They might even kill me."

"Tell the guards this. They cannot send you back."

"I have told them and begged them, with a lawyer. There are such lawyers here who would help me." He sighed. "And you'd think after all the help I give the guards they'd do something for me, wouldn't you? But no. They laugh, and pretend like I should be packing for a long holiday."

He wept tears of fear, not self-pity.

"I'm sorry, Uncle."

He sat up on the bunk and offered me a dignified handshake. "Thank you. So am I."

The next morning I watched as the guards led Bernard out of the cell block. He didn't look at me. He just stared at the ground mumbling to himself.

I don't know what happened to him, or if he is a free man today. Maybe he's dead. I wish I knew, because I think of him often. There was something about him that reminded me of my father.

I BEGAN TO MAKE PLANS—I WAS NOT GOING TO END UP LIKE Bernard: I had to get out, to survive. The authorities could not be trusted any more than the people smugglers. They were all liars, and were only looking to take advantage of the human river that ran beneath their feet. My greatest fear was to be sent back home: it would have felt like such failure. As much as I longed to see my mother and family, I did not want to shame myself or them. As hard as my journey was, they had invested everything in my success.

I thought maybe I could try again: I could walk to Afghanistan's western border with Iran, like my old friend Shah had done. Maybe I could even still get to Europe and find Hazrat. I could find a job and save money. Find Black Wolf—maybe he would help me? My mind was racing, none of my thoughts realistic.

Several of the Afghan prisoners who had arrived in jail at the same time as me had already been given the paperwork confirming their deportations, but there was a catch: the Afghan embassy would help arrange their flights for them, but they had to pay for it themselves. If they couldn't, they stayed in prison. There was a single pay phone near the toilets. From my sleeping space so close to the cell doorway, I could hear snippets of conversation as imprisoned men apologized to families already in debt to people smugglers, and asked them to find the money to pay for an international flight. Listening to them talking to loved ones was like a double stab through my heart: their plight could so easily be my own—and yet I had no phone numbers or way of contacting my family.

Some of the men wept like babies as they came back. It was through this I learned that people smugglers don't pay refunds for failure. If you get caught and finish the game early, you still have to pay in full. The agents will normally offer a discounted deal to let you try for a second time, but if you don't take that deal, your family's money is forfeited.

As yet, I hadn't been given any papers, so I had some hope that things might work out.

What I did have, however, was a piece of paper with a phone number for Zamir, the flashy young Afghan who had been our second to last agent. Abdul suggested we try calling him for help.

Mehran and I stood by the phone expectantly as Abdul spoke to him. We could see from his face it wasn't going well.

"What did he say?" Mehran asked impatiently, as Abdul hung up.

"He told me not to call him from prison. He was worried the police would trace his phone. Said it was our fault we got arrested."

We had no way of contacting Qubat, the so-called big guy in Kabul.

Abdul had asked Zamir to at least call him on our behalf and let him know where we were, but Zamir had said it wasn't his responsibility before telling Abdul never to call him again and hanging up.

We also tried Malik, the be-suited very first agent in Turkey, the one who had made us live in the chicken shed. But the number we had for him was now unobtainable.

I think I had been in prison for around two or three weeks—time dragged so slowly it was hard to know for sure—when the man from the embassy came again.

He explained that those of us who couldn't afford flights were to be deported to Iran by coach the next day.

JUST AFTER BREAKFAST THE FOLLOWING DAY, THE GUARDS came to get us. "You men on the bus to Iran. Get ready. You have two minutes."

I jumped up in a panic, as if I had lots to pack. Then I remembered I only had my small bag, containing all the possessions I had. "I'm ready."

Before we left, Mehran, Abdul, and I tried to ask the guards about Baryalai, saying we didn't want to go without him.

"Shut up, you'll make them suspicious. It will make it worse for him," another prisoner warned us.

Leaving without him and not knowing whether he was still in a different prison somewhere, or had already been deported, was horrible. But what choice did we have but to go?

Yet again I was going backward on this twisted board game that had become my life—but part of me was happy, because at least I was getting out of prison. The confinement had been beginning to make me lose my mind.

CHAPTER 14

After four days, I was beginning to wish I was back in prison. Around fifty of us were packed into a large coach, two to a small seat, our sweat and breath intermingling. I sat next to Mehran. Every six hours we pulled over to go to the toilet. They gave us very little water, so it was hard to go.

I had lost all hope. I was going back to war and back to Afghanistan—where the Taliban would be waiting for me.

Some of the men started grumbling.

"There are only four of them, and fifty of us. We can take them easily."

"At the next stop, five of us can grab each one—another five the driver."

Maybe this sort of dissent was common at this point in the journey, because the guards, plain-clothed immigration officers with guns, seemed to know what was going on.

A guard pushed his way to the back of the bus, an anxious look on his face. "We're almost there. Don't get any clever ideas. Things can always get worse."

He didn't even bother to pat his gun. He didn't have to. We'd all heard the stories from other prisoners. It was alleged that sometimes Turkish border guards sold refugees to the Kurdish criminal gangs. These gangs, which everyone referred to as mafia, would imprison you somewhere and then make you contact your family to demand they pay a ransom. We were told they always asked for a ridiculously high price, like $5,000, which you had to try to negotiate down. If you didn't cooperate and refused to call your family, they made you walk barefoot into the mountains, where they beat you. In extreme cases, I'd heard of men having their noses or ears cut off because their relatives either couldn't or wouldn't pay. As ever, the ransoms were paid to a middle man with a bank account somewhere in Europe or Iran. Nothing could be linked back to the kidnappers directly.

For me at least, this meant I was on my best behavior with the immigration officers. I didn't want to be sold on like that. Aside from Black Wolf, my experiences of Kurdish smugglers and criminals so far had not been positive.

On the fourth day, when we were very close to the border, we were ordered off the bus to sit and wait for a few hours in what looked like some abandoned former military buildings. The whole area was covered with tiny bits of sharp gravel. It was too tiring to stand up all afternoon but it was painful to sit on, and I was like a chicken on a hot plate, constantly shifting and fidgeting as the gravel pricked my bottom and thighs.

As we boarded the bus, the count came back one person short.

"Who is missing?"

Somehow one of us had managed to disappear. I looked around, but it was hard to tell who it was.

The guards stayed with us on the now stationary bus, but they informed the police in the local area.

Three hours later, the police returned with a guy called Zabi. He was also a Pashtun, from the city of Kandahar, and only a couple of years older than I was. He was bleeding and bruised, but still had some fight left in him: "Let me go, you bastards."

The local police who had caught him shoved him hard into the side of the bus. "You stupid fool. Did you really think you could run away from us?"

He carried on screaming insults at them, which one of them took as justification to punch him in the guts.

Satisfied, they shook hands with our guards and drove away in their blue and white Hilux.

Zabi told us his story: "When I went to pee, I just ran. I walked for ages. People kept looking at me but I carried on. I was trying to find the bus station."

Everyone had questions for him.

"Did you have money?"

"Why didn't you ask me too? I would have run with you."

"You fool. Now you've put us all at risk. They will be angry and sell us to the mafia."

The guards weren't happy. "Shut up. Don't talk to him."

It was late when we got to the actual border. We didn't go to one of the main border crossings; this one was just some kind of a military checkpoint. There was no traffic—a single man on a motorcycle droned through like a blowfly. Other than the half dozen Turkish soldiers standing around, there were no people there.

The guards came toward the bus, carrying a large box of *zataar*—hot Turkish flatbread sprinkled with fresh thyme. It smelled delicious. "We can give you this now, or you can eat it before you cross. Your choice, but we suggest you eat it later."

In my world "later" could mean anything, but the general consensus was to go with the guards' advice. Bad idea, because we never did get to eat it.

We were put into a series of army jeeps and driven along a road, down one side of which ran a very high barbed-wire and electric fence. Every so often along it, there were lookout posts with bored-looking soldiers inside. We were told to get out of the jeeps and wait by the fence. Suddenly, a lot of chatter and noise came through our guards' radios, and the soldiers who had been driving the jeeps rushed to the wire, unclipping a long strip of electric fencing. They told us to wait but get ready.

It was dark by now, made more so by heavy clouds that obscured the stars and moon. Then the soldiers turned the jeep headlights off, making it even scarier.

Mehran turned to Abdul and me. "Why would they do that? Do you think the Iranians know we are coming?"

Abdul looked tight-lipped and grim. "Maybe they know the mafia are there."

There was one man on the bus who spoke fluent Turkish, an elderly Pakistani man from Peshawar. He'd spent most of the bus journey chatting happily to the immigration officers. They clearly liked him and, I suspect, felt a bit sorry for him. As we prepared to move, one of them handed him a large knife in a leather sheaf.

"Here. You take this. Be safe."

It was a nice act but it didn't make the rest of us feel any better. They clearly knew full well they were sending us into danger.

"Go. Go. Go."

When the order came to walk, Mehran held me back.

"Don't be first across."

A few men had already started to walk through the gap in the barbed wire into Iran.

I had no idea what was waiting for us out there in the dark. My legs were like jelly.

"Go. Move."

It was all I could do to make my feet move. Mehran took one of my hands and Abdul the other. I was slightly reassured to see the old Pakistani with the knife was right next to the three of us.

Everyone by now had started walking. We were in small groups but in a disciplined horizontal line, like soldiers coming out of the trenches in the First World War into no-man's land.

People walked slowly at first, gingerly, carefully. Then suddenly they started running, going in all directions.

We broke into a run with them.

Then we saw some police waving at us. "Come this way. Over here." There were eight of them, all carrying horsewhips.

We started to run toward them, but then we heard other voices, shouting at us. "No. Not that way. We are the police. Come here. Over here."

"It's the mafia," a frightened voice shouted. "The Kurds. They're in police uniforms. Run."

"Where do we go?" I was panting both from fear and from the running—the weeks in the prison had left me exhausted and weak.

As everyone ran in panic, I lost hold of the others' hands as we were knocked and buffeted, swept along with the frightened herd. The ground was rough and rocky, and I thought I would fall and be trampled. My chest was hurting so much I couldn't catch my breath.

Headlights appeared, then a few men wearing uniforms and carrying torches walked toward us. Suddenly there were more cars, trucks, motorbikes. The men with the horsewhips were upon us, their whips making cracking, splintering sounds as they struck at the running bodies. They managed to corral maybe half of us into a semicircle.

"We were informed you were coming. Stay calm. Do not move."

Some people tried to break free. A whip cracked through the air.

"Do not run. Do not run."

I had no idea who these men were but I truly hoped they really were the police. One of the motorbikes moved round behind us, as the half dozen men with the whips ordered us to walk forward.

As we walked on, dawn began to break and it was possible to see a little more clearly. Up ahead I could make out a one-story building. There was an Iranian flag flying above it, and soldiers standing outside.

My breathing began to return to normal. We were with the right men.

As we approached the building, I realized for the first time how cold it was. We had been in a stuffy bus for days, but now on this flat, rocky plain with spiky bushes and mountains ringing the distant landscape, I shivered. It was definitely winter, maybe February by now. It could even be March. I really had no idea.

The one-story building appeared to be the first Iranian checkpoint on this side of the barbed-wire border. Watchtowers stood to either side of it. It was still semi-dark, and moths fluttered around a fluorescent light above the doorway. As we arrived, three more police officers spilled outside.

"Line up. Now. Hands on your heads. Faces to the wall."

We placed our hands on the front of the building.

"Slowly turn around. Empty your pockets, take everything out."

They spoke to us in Farsi but their accents were heavily Kurdish. I guessed we'd crossed the border back into another Kurdish-dominated area.

Some people were not cooperating with the order and didn't want to empty their pockets. As a result of that, we were all ordered to take off our shirts, even though it was freezing cold.

The Pakistani man had tried to hide his knife but he was rumbled.

"Where did you get this?"

"In Turkey. The police gave it to me."

"Liar." The slap rang out clear in the crisp air.

Why was it every time I thought I was about to feel safer, the opposite happened?

I had $250 still: $50 was in my bag; $100 was hidden in my underwear; and another $80 was in my socks.

From my pocket I took out a $20 bill.

A policeman looked at it, took it, pushed me, then put it in his pocket.

His friend began rifling through my bag. "Got any drugs or guns in here?"

The policeman who had taken the $20 laughed. "Don't you know it's good manners to bring gifts when visiting others?"

I was furious. We might as well have been taken by the criminals.

"Look at this," his friend exclaimed. "What a little discovery. Green's my favorite color."

My heart sank.

He scrunched the $50 note like an autumn leaf. It was the very same note I had had stolen from me in prison, the note that had only been returned to me because Congolese Marrion had beaten another man on my behalf.

I knew I wasn't getting it back, but this time I didn't care. It felt toxic to me now. Maybe this was God's punishment for what had happened.

And at least my $200 was safe in my pants and socks.

We were still standing there, freezing and half naked, when a series of cars arrived and parked by the entrance. Several men jumped out. The police officers walked over to them as if they were old friends. All of them stood by the cars, speaking in rapid-fire Kurdish as they argued and negotiated something.

"They are bloody plotting to sell us. These guys are mafia," someone whispered.

My Kurdish was good enough by now to pick up a few awful words.

". . . depends how much . . ."

"Let them rot in . . ."

". . . risk . . ."

"Families pay . . ."

"Your share . . ."

I felt sick with fear.

Mehran was trying to reassure me. "It doesn't make sense. If they wanted to sell us, why did they bother to come and save us?"

"Money," said Abdul. "Why does anybody do anything to us? We are just another dollar for these people."

I wanted to run: "We have to escape." But even as I said it, I knew there was no way to escape. They all had guns and I was not physically strong enough to risk anything stupid. I'd barely eaten anything in days.

Abdul read my mind: "How? Even if we could get out of here, do we just run into the hills? No water, no food? We would be dead in two days—even if they didn't just capture us again."

Just then, the low hum of an approaching vehicle could be heard. The voices at the entrance changed, becoming more urgent.

I tensed. This was bad.

A young man with epaulettes on his shoulders got out of his car. He looked like a boss. His smart officer's uniform strained around his belly as he shouted at the other policemen.

He spoke in Farsi: "You think I don't know what you are doing? Meeting your gangster friends in the middle of the night. This lot"—he waved his hand in our direction—"like chickens ready for market."

My legs started to shake with relief.

"Tell your other friends to leave. If I see them here again, they too will be in the cells. Is my order clear?"

With that he strode inside. The men in the cars left, but only after handshakes, hugs, and apologetic looks from their police-officer friends.

We stood waiting by the wall for over an hour in the cold, but at least we were allowed to put our shirts back on.

I think the boss must have called people he knew personally to take us, because different cars came. Not police cars, but what looked like local people on their way to market. We were ordered into them. A couple of the police officers came with us.

The car I got into had a trussed-up goat in the back. It bleated at me pleadingly as we drove away. I knew just how it felt.

CHAPTER 15

THE MOUNTAINS.

After a while they start to look like bones. Jagged knuckles, bare of flesh.

I felt as though all life had been stripped from me.

My face felt like granite, sharp and hard—washed clean by storm after storm. I tried to sleep—I was exhausted but, as ever, I was the smallest person in an overcrowded car.

From the border we had been taken to an Iranian town called Maku, where we spent the night in a local police station. We still hadn't eaten and I was faint with hunger.

In the morning we were put into police cars and told we were going to court.

The land around Maku was crisscrossed with apricot orchards and vineyards, watered by a network of irrigation canals.

"It is very ancient here," said the police officer in the back. "It is very beautiful, do you not think?"

I didn't care. I ignored him.

We drove through a gate in a high stone wall.

"Look," the driver spoke. "Look at this tower."

It was. A seven-sided tower loomed, ornate and strong; maybe a grave for a whole dynasty.

"Maku is famous for its architecture, as you will see." Then the driver laughed until he coughed.

They took us to a golden-brown-colored court building—it made me think of bread. My knees wobbled as we were led inside. I didn't think I could face another interrogation like the one in Turkey.

Many of the court officials and local police spoke only in Kurdish, not Farsi. We were given documents to read in Farsi, but the language was so complicated I couldn't really work it out. I was just about to ask Abdul to try to explain it to me when we were told to get up and walk over to a counter. There, we were told to write our names, ages, and countries of origin in a big blue ledger.

I wrote my age, twelve, in big clear letters. I hoped they might take notice of it.

The whole court process was over in a few minutes. Escorted by the police, we were made to walk from there to a nearby prison. There, the police did some kind of official handover to the prison officials, showing them the court papers. Abdul said he thought the court papers were permission to detain us. I prayed someone would notice my age and do something to help me.

The prison was some kind of old military barracks made of mud and stone. Now I understood why the driver had laughed so hard. The barracks were indeed ancient, with no running water. We had to file under a carved archway and down a little step to get into the dungeon-like cells. On the border it had been so very cold, but now it felt so hot in the airless cell I was assigned to, sharing with ten others. There were no beds, only mattresses on the floor, packed in such a tight row that you had to step over them to reach the ones closest to the far wall.

There were about twenty-five prisoners, all refugees, already there.

"Welcome to the luxury hotel."

It was what was by now becoming a familiar story: they told us if you couldn't afford to pay for your own deportation, you rotted in that jail. Some of them had been there weeks, some months.

The prison warden was a fat man with a uniform that looked fit to burst whenever he moved. To cheer myself up, I imagined how funny it would be if it did, his buttons popping everywhere.

He issued us our instructions: "During the day you will be outside exercising or resting in the yard. At night you stay in your cells. If you want to get out of here, you can pay from amongst yourselves for the cost of a coach to the city of Zahedan, on the border with Afghanistan. The authorities will help return you to the hands of your government. The Pakistanis among you can go home from there."

This caused consternation among the Pakistanis of the group. Afghanistan was not a safe place for them, especially with no papers.

"There is no food for you here. If you are hungry, give us money, and the guards will go and get it for you."

The first thing we did was pool some resources and send out for some chicken and rice. I was delighted to finally get something to eat.

Each morning, the guards turned us out into the yard. There was no shade, and this fact was made worse by hundreds of dive-bombing mosquitoes, which filled in the gaps of misery the sun could not reach.

The guards sat and watched us bake from the shade of their watchtowers. They made us do endlessly humiliating squat jumps with our hands on our heads. My head swam in the heat. Each breath scorched my lungs so painfully that I felt my whole body might burst into flames. My arms and thighs sang from the pain, but they kept forcing us to do more and more until we were on the point of collapse. Only then did they let us rest.

We had little food the week we stayed there. I had money and so did Mehran and Abdul, but we couldn't afford to use all our cash to buy food for everybody. As we felt guilty eating in front of hungry people, we mostly stayed hungry ourselves. One afternoon, after the forced exercise, a guard threw a crust of bread—the remains of his lunch—into the yard. Two desperate men jumped on it.

I knew I had to get out of there as fast as I could. I took it upon myself to try to find enough people with money to help pay for the coach to Zahedan. The governor had made clear that the coach to freedom was only permissible if it was full; the journey to Zahedan was far, and he would need to send guards with us. He said it would cost us $20 per person.

A man named Raheem and I began to negotiate with him. In the end, the governor agreed on $10 per person.

As there were fifty-two of us in total, that made $520 for him, minus the bus fee, of course. But there was one condition: "You take everyone. I cannot listen to the complaining a moment longer."

With Raheem's help, I became the organizer. As a child, people were more trusting of me not to cheat them. I walked around the yard trying to get people to cough up. Some were genuinely penniless, begging me to help them; others had money, but wouldn't admit to it. Some were willing to pay for themselves and a friend only. Others refused to pay at all.

I had to use all of my powers of persuasion.

By the end of it, I was $50 short of the $520 we needed. I used my own money to make it up. That left me with just $150 to my name.

It was worth it, though, the moment we boarded that coach.

FROM MAKU, THE COACH HEADED TO THE CITY OF SHIRAZ; from there it would go on to Zahedan and the border with Afghanistan, from where we would be deported.

It was a very long way but I didn't care—at least I was out of that prison. The bus was ancient, fiercely hot with no air-conditioning, but the windows opened and we were in the shade most of the time. Two prison guards were assigned to travel with us.

"Behave and cooperate, and we will treat you with respect," they told us curtly.

Every now and then we stopped at a small town along the way for food and toilet breaks. A guard went around collecting money and returned shortly with tea and bread for all of us. I was touched when one of them paid for our food himself on one occasion.

They usually let us get out of the bus when they were eating themselves. I think they felt embarrassed to be eating in front of us because they had nicer food than we did, and they knew we were hungry.

We had stopped to rest in some small public gardens. We stretched out in the shade of a low tree and ate hungrily. As usual, Mehran, Abdul and I stuck close to each other.

As we sat chewing in silence, we all looked at each other. The same thought—escape—passed between us unspoken.

"If we are going to do this, then we must do it right," I said, knowing that failure meant the police would probably shoot us.

"It is madness," said Abdul. "Don't be so stupid as to even try."

I looked at Mehran.

He nodded to me. "I'm with you."

I can't remember how many days we had been on the road by then. I think maybe three days and two nights. We had a feeling this could be our last stop before we arrived at Shiraz, so this was as good a chance as we would get.

"Raheem is a brilliant Farsi speaker. He can help us get across Iran."

I looked over at Raheem. He nodded in agreement.

"I'm not helping you get yourselves killed," said Abdul.

"Come on, man. Let's stick together," Mehran pleaded with him.

"Nor me," said the old Pakistani man, the one who had been given the knife in Turkey. "I am too sick to run. But I can distract the guards for you."

"But that's the point—they don't expect us to run. They think we are far too exhausted and hungry," said Raheem.

"And that," I said, "is exactly why we must run. We must run now."

CHAPTER 16

FAR SCARIER THAN THE POLICE AND THEIR GUNS, WAS MY mother and her disappointed face. For the first time since I had set off on this journey, I knew exactly what I needed to do.

I really didn't want to be deported, but I could sense the others were unsure. After all, I was a twelve-year-old trying to persuade grown men to escape a prison bus with me.

I knew I had to try to persuade them that this was a risk we needed to take; I tried by pushing the most obvious button you can with Afghan males: by calling their bravery into question.

"If I am not scared, why are you? Isn't it better to try than regret?"

Mehran nodded enthusiastically. He was as keen as I was.

No one else said a word. A couple of people looked at their feet or picked at the grass.

Abdul was still adamantly against it. "Are you crazy? You'll be shot."

Mehran sat quietly for a few seconds before weighing up his options: "If they send me home to Afghanistan, I am dead. If they keep me starving in one of their prisons, I am dead. Europe is the only chance I have."

Our kindly decoy, the old Pakistani, climbed to his feet, brushing his trousers clean. "May Allah be with you. Are you ready, boys?"

"Are you sure?" I asked him. We all knew he was risking punishment for us.

"Yes. I'll try to make sure you get a head start."

"How will we know?"

"Don't worry, you'll know. Just run as soon as I reach the bus."

He walked toward the parked bus, screened from our view by a bed of tall shrubs. The prison officers were sitting on the other side of it, eating. We had all behaved so well for the past few days that they had begun to relax and let their guard down a bit more than they should have done.

And then it happened. "Guards," our old friend cried. "A dog just stole my bread. Tried to attack me."

"What are you talking about, old man?"

"I think he might have rabies, sir."

"Rabies?" said the officer. "Show me where it went."

And the old Pakistani gentleman led the guards in the other direction.

We seized our chance.

I looked around at the nervous faces. Abdul shook his head. Raheem stood up.

"Let's go."

Mehran and I both sprang to our feet and ran through the park with Raheem following right behind us. We dashed out of it and into a narrow side street.

My one abiding regret is that it all happened too fast to say a proper good-bye to Abdul—until the very last second, I don't think he believed we were really going to do it.

Until we ran, I wasn't sure I had believed it myself.

WE RAN THROUGH A SERIES OF NARROW ALLEYS LINED WITH tightly packed houses, trying to avoid the main streets where there was more traffic, and a group of sprinting, breathless, wild-eyed males would be quickly noticed.

It wasn't a big town, but there was much confusion about which way to go. There were fields ringing the town, which we'd seen from the bus window, but we had no idea how to get there. I half expected to feel a bullet in my back at any moment.

I was pretty confident the prison officers would not come after us because they had to stay with the rest of the group, but I knew they'd call the local police, who could be waiting for us around any corner. Suddenly, I realized that the houses were giving way to shops and there were more people around.

We were running into the center of town, not out of it.

"Wrong way," Mehran panted breathlessly, his eyes wide with the dawning realization of what we'd done. "Where now?"

"Look normal," Raheem murmured to us both.

We slowed to a walk, trying to look as casual as possible but it was hard, and my heart was pumping with fear. It was clear we were distressed, dirty, and scared—we looked like the escaped convicts we were.

Quick-thinking Raheem asked a passerby where we could find a taxi. Luckily, we got one almost straight away and, to our joy, the driver was an Afghan immigrant.

"Please take us to Shiraz."

I think he guessed something was up, possibly even realized we were illegal refugees. He reassured us with a single sentence. "I understand; you are lucky to have found me." Shiraz was also where the prison bus was headed for an overnight stop. But we knew it was a big city with a lot of non-Iranians, especially Afghans, wandering around. It would be easy for us to blend in and avoid detection, something impossible to do in the current town we were in. We had to get away from it fast if we stood any chance at all.

I breathed a sigh of relief as I sank into the black vinyl backseat of the taxi with Mehran.

Raheem got into the front and spoke to the driver. "*Kaka,* we don't have any Iranian money. We only have U.S. dollars."

The driver looked slightly frustrated but was still kind: "Give me your dollars. I will exchange them for you."

Raheem gave him $250.

At the bus station in Shiraz, the driver exchanged Raheem's money at a little kiosk. As he drove off, he offered us some sound advice: "Don't get lost again. You might not be so lucky twice."

The station was buzzing with people. We were able to relax a little bit because there were refugees everywhere. Shiraz is a little bit like an Iranian version of Dubai, with lots of people from neighboring countries going there to work.

We weren't really sure what to do next.

"Do you know anyone in Iran?" I asked Raheem.

"Not anyone I can call. You?"

"No."

"Yes, we do," said Mehran. "What about the place we stayed in when we first came here, in Tehran? Where we met Shah and Faizal. The place was run by that nasty guy—the one who took our passports."

"Why would we call him?" I still had bad memories of that guesthouse, which stank of boiled meat.

Remarkably, Mehran had the phone number: he had meticulously stored all of the different agents' numbers in a notebook in his backpack. I was amazed—he was the joker of our pack; he'd never struck me as particularly organized, or even that intelligent.

I didn't think for a second it would work. It had been several months ago that we had stayed there, at the end of October 2006. Now it was late spring in 2007. I didn't think the man would even remember us.

The only other place we had stayed in this country, aside from the Maku prison, was the smart hotel in Mashhad when we had first arrived from Afghanistan last October. Mehran had written down the telephone number for that too, but not the name of it. Besides, Mashhad was at the opposite side of Iran, to Shiraz. If we were to avoid deportation and make it to Europe, we needed to get away from border cities and to the north of Iran, to Tehran.

I stood lookout with Raheem as Mehran made the call from a public phone booth. He spoke as loudly as he dared to make himself understood over the noise of the street. "We are people of Qubat." His eyes darted around nervously as he spoke. "Three of us. Escaped from the prison bus. We need help."

I shifted anxiously and scanned the street for police or anybody that might be taking an interest in us. It was the early evening rush hour, and tired commuters were busy rushing home from work or running for buses. Thankfully, no one took notice of us.

After Mehran hung up, we walked to a quiet corner where he repeated the conversation to me and Raheem: "He said we have two options. If we go there, he'll send a friend to get us. He says he is too busy to help us himself. Or he says he knows someone in Shiraz we can stay with, but he said that would be more risky."

"I don't trust a word he says," I said. I hated that man.

"Me neither, but he's all we have," continued Mehran. "The good news is that he told me he'd call Kabul and inform Qubat of the bad news that we are back in Iran. He said it was up to Qubat to decide what the next plan was. If Qubat agrees to pay him, he can help us himself, and we can go to stay at his place again."

"I will look forward to that," I said sarcastically. "Maybe we can get our passports back."

"I think they were sold for a tidy profit long ago, my friend."

The day was fast disappearing into a warm and sultry night. We used some of Raheem's newly exchanged Iranian money to buy tickets for an overnight sleeper coach to Tehran. It wasn't leaving until midnight, some six hours away.

We were absolutely filthy. We had spent a week in the prison at Maku, and three days on the hot and sticky bus journey, all in the same set of clothes. I worked out that I hadn't had a shower since the prison cell in Istanbul, about fifteen days ago.

Raheem was worried that the state of us might arouse suspicion. I got his point. Mehran and I sat on a bench and waited while Raheem wandered across the street to a small shopping mall. He came back with three new shirts and trousers for us all, as well as socks and underwear. With a look of triumph, he then produced a small glass bottle from a plastic bag. "We can't shower, but I brought some cologne. It might help."

We managed a quick wash as we changed in the public restrooms. The cologne, a cheap musky sandalwood, might have made us smell stronger but it certainly didn't make us smell any sweeter. But I felt quite grown up as I looked in the mirror and copied Raheem, who was splashing it all over his neck and behind his ears. I had never worn cologne or aftershave before.

We threw our old clothes in the bin. It was so good to wear new clothes—but Raheem had managed to bring me jeans one size too big. I didn't have a belt, so I kept having to hook them up as I walked.

"You two look like Bollywood stars—especially you, Gulwali." With that, Mehran did a silly dance of the type we'd seen in the Indian DVDs we'd bought.

My face burned red. I was a bit sensitive about the fact that I had watched and enjoyed the movies back in Istanbul. For me, it still felt a bit *haram,* forbidden. Remembering the films also made me think of Baryalai, who we had left behind in a Turkish prison, accused of being an agent.

Feeling more secure in our new clothes, we ventured out of the bus station to try to find a restaurant. The last time I had had proper food had been back in Istanbul, before I went to prison there. That had been well over a month ago. In that time, I'd been to two different jails in two different countries and survived on either tiny amounts of rations or nothing.

To our delight, we found a tiny Afghan restaurant tucked away behind the station. It looked just like the restaurants in Jalalabad used to: Formica tables, metal chairs, plastic roses in vases, and mirrored tiles on the wall in angular patterns. For the first time since leaving home, we had proper Afghan food: Kabuli *pilau* (rice fried with onions and raisins); kebab and *bolani* (a type of fried pasty); all with freshly baked naan, washed down with salty *lassi* (yogurt) with cucumber and mint.

The scent of the deliciously cooked food was the smell of home. It was the type of food my mother cooked so well.

I so badly wanted to enjoy this rare treat, but my stomach had shrunk and was painful from the previous lack of food. I had to force down every mouthful, and eating it made me feel instantly bloated. I was so disappointed, and soon in a lot of pain. We wrapped some of the kebab in napkins to take with us for the journey.

THE BUS JOURNEY TO TEHRAN FROM SHIRAZ TOOK ABOUT eight hours. I'd never traveled in a coach so luxurious. We were thrilled to discover that this one was a proper sleeper, with air-conditioning and seats that reclined into beds. It even had a TV screen above the driver's seat showing movies. To me, it was like a mini-hotel on wheels.

I slept soundly, only waking when we stopped in Esfahan for breakfast. Esfahan is famous for its architecture, with many beautiful palaces, covered bridges, mosques, and minarets. There is a famous saying about it, "*Esfahân nesf-e jahân ast.*" ("Esfahan is half of the world.") It was a place I had seen in picture books as a child, and while I was scared of being arrested any second, I did still enjoy seeing this wonderful place.

When we reached Tehran, we called the number Mehran had been given for the man who was supposed to come and pick us up. I was so scared that the number might be a hoax, and that no such man existed. But, thankfully, a man did turn up.

He was Afghan and said he worked in Tehran legally, laboring in a warehouse. I got the impression he wasn't used to dealing in this refugee business—I think he'd been co-opted into helping us by the guesthouse owner.

He took us on the metro. This was truly exciting: I had never seen such a magical thing before. The carriage was made of glass, and I was amazed by how it was we were standing in the same place but the carriage was moving. It was beautiful.

After half an hour on a couple of different trains, we arrived at the other side of Tehran. The man walked with us to a local park, where he left us and told us to spend the day. He said he needed to go to work and would come and pick us up in the evening.

We spent the day trying to relax but it wasn't easy. Every time someone walked past us, we were nervous because we thought they were staring at us. I felt as if I had the words "Escaped convict" tattooed on my forehead. Fortunately, Raheem still had some Iranian currency left so we were able to buy drinks and snacks to keep us going. My stomach was still not used to normal food, though.

When the man returned in the evening, as promised, he took us to his house. It was a very basic one-room hut on a scruffy farm on the outskirts of the city. He explained that the farm belonged to an Iranian friend of his and that he was allowed to stay there in return for keeping an eye on the place when the man was away.

We ate and slept on the hut's flat roof. It was the only place where there was enough room for four people, but it had the added advantage of offering us a good vantage point should police cars come looking for us.

Our benefactor was, as we had suspected, neither a paid smuggler nor an agent: he was helping us as a favor to his friend. I don't know if he'd been put under pressure to do so, but he was very gracious about it and we were grateful to him for putting himself at risk to shelter us.

He told us most Afghans living there, even the ones who were legally allowed to work, lived in poverty and struggled to survive. "It's tough, but it's better than home. At least we are free from bombs."

In the morning his friend, the guesthouse manager, called and asked him to let us stay another few days. It seems that he had contacted Qubat, who refused to pay him for the intervention or hire him to move us on again. Qubat wanted us to go to one of his different—we assumed more efficient—agents. But this guy was refusing to let us go. He insisted that because we'd called him from Shiraz we were now de-facto "his" for as long as we were in Tehran. He insisted Qubat owed him money, at least for the help he'd given us so far.

In our minds, if anyone deserved money it was the man whose hut we were sitting in, not the guesthouse owner. But, of course, our host had no direct connection to Qubat himself—he was just a poor Afghan caught up in the middle of all this.

The whole thing got even more confusing when two other refugees arrived. They had only just left Afghanistan and this was the first step in their journey to Europe. Jawad, in his forties, was from Nangarhar, the same province Mehran and I came from. Because of that, we nicknamed him "cousin."

Jawad was very sad. Usually, people didn't say too much about their background stories or personal reasons for leaving, and he didn't tell us the full whys and wherefores. But he often spoke of his little son, who was just six. He was missing him and his wife like mad.

"Will I see them again?" he asked. "I can't bear it."

The other person was Tamim. He was very young but had lost all his hair due to stress, although he liked to joke it was because, "I think too much." He was a tailor from Jalalabad. Somehow, he and Mehran worked out they were distant relatives by marriage. They decided to call each other "cousin" too, which meant overnight Mehran and I suddenly had not one but two new relatives.

Tamim told us he had been threatened by the Taliban and that's why he'd left. Why exactly, I don't know.

On the fourth day it seemed the argument with Qubat had been settled because our host was instructed to take us to the side of the field running alongside the farm, later that night, when it was dark. A deal had been struck with a bus conductor, who would stop to pick us up.

As ours wasn't a proper bus stop, we rather assumed the bus would be empty, but when it arrived it was packed with locals. We were fuming. How obvious could it be this was a dodgy pickup? We were certain we'd be caught.

Raheem stayed behind. Our host had reassured him that, even without papers, he could work in Iran. He knew enough about the country and had such good language skills he figured he'd be better off trying to survive there than continuing to risk all to get to Europe. I was sad to leave him, but I understood his reasons.

The new group of four: me and Mehran, and now Jawad and Tamim too, traveled on the bus for the whole night, until we were the only passengers left on board. We had no idea where we were going, only that we were driving away from Tehran. I hoped, prayed, that the bus was going somewhere in the direction of the Turkish border, which was what we needed to do in order to keep moving forward.

If I made it, it would now be the third time I had crossed the border into Turkey.

CHAPTER 17

"GET OUT."

It was ten o'clock in the morning. The driver had just stopped the bus, without warning.

"What? Where are we?"

He was dropping us in the middle of the road, in the middle of nowhere.

"Get out."

We did as we were told.

"Now what?" Jawad and Tamim looked at Mehran as the bus drove off.

He shrugged.

"Have you got a plan?" They turned to me next.

"Nope."

"Oh, that's just great," snapped Jawad. These two were new at this. They had yet to realize that this kind of stuff was normal.

Not knowing where I was, where I was going, or when I'd get there was normal for me now. Dealing with constant uncertainty was not only depressing, it left me in a state of permanent adrenaline-fueled anxiety.

We stood by the road not knowing what to do. Less than five minutes later, a car approached us.

It was the most amazing car I had ever seen. It was silver, with tinted windows and looked like a racing car, of the type I had only ever seen in pictures. As the driver pulled over, his car sat crouched low to the broken ground, like a wild animal ready to pounce.

A young, clean-shaven man in jeans, a gray Western-style suit jacket, and a red shirt opened the door. He looked very nervous as he spoke: "Qubat's people?"

I couldn't contain my smile. It was so wide I thought my face might crack—even as hungry, tired, and scared as I was, that car gave me a thrill. If my friends back home could see me now. Imagine driving through the streets of Jalalabad in this.

"What are you grinning at?" said the man, who was dabbing sweat from his brow with a handkerchief.

"Nothing," I replied, swallowing my teeth.

He looked really annoyed. "Why are there four? There should only be three of you. I can't take four." He was right. The car had only one passenger seat and a very tiny backseat. He pointed at Tamim. "You wait here. I'll come back for you."

"No way. I am not staying."

We looked at Tamim, feeling guilty. I looked at the man pleadingly: "We can try."

Two minutes later, Mehran and I were crushed into the tiny backseat, Jawad, who was the eldest, was in the front—even in these strange circumstances we always gave our elders respect—and Tamim was in the trunk.

The man was clearly not happy—sweat was now trickling down his temples. He didn't take us very far: along a couple of winding lanes and into a small hamlet, eventually pulling into the yard of a small brick house. In front of it was a woman sweeping, using one of the stick brooms my mother used.

The man ordered us out of the car. "This is my parents' house. You"—he pointed to Tamim—"will stay here, and I will come back for you."

Tamim started arguing: "Why me? Leave one of the others."

I felt bad for him but I wasn't about to give up my place for him. I'd been on this journey for over half a year now; Tamim had been on it for less than a week.

Mehran and I got in the tiny backseat again.

The engine roared into life as we pulled away onto some small and winding but smooth country roads. Tamim looked forlornly on from the yard. Mehran and I grinned in silence at each other as we pulled away.

After about an hour, the driver pulled over again. "We are nearing a place where there are lots of police searches and checkpoints," he said. "I need to take some precautions."

We all climbed out, and the driver opened the trunk. Jawad and Mehran looked at me expectantly. I moved toward the trunk: as the smallest and youngest in every situation, I accepted the treatment.

"No," said the driver. "The boy rides with me. You two—get in."

My friends stared in shock at the tiny space.

"It's only a few miles. Then you can get out."

Neither of them moved.

"Look, the boy can pass as my young brother. You two look like a pair of illegals, which is what you are. So, what do you suggest we do?"

I tried very hard not to burst out laughing.

Jawad went in legs first, tucking his back tightly to the back seats. Mehran followed—just: Jawad had to hold him in place by wrapping his arms around his chest so that they neatly spooned each other. I don't know how two grown men managed to fit in there, and it did not look at all comfortable.

The driver smirked. "Such a cute couple."

I could barely contain myself at that—I bit down hard on my lip and snuffled out my nose.

Mehran looked up at me, his face red with discomfort and humiliation. "You laugh one more time, Gulwali, and you'll pay for it when I get out of here."

"Mind your heads," said the driver, shoving the trunk lid closed.

Barely able to contain my glee, I climbed into the passenger seat.

"Put your seat belt on," the driver barked at me. "We are relatives. So you must call me 'brother' if the moment requires it."

"Yes," I said, flushing with guilt, anger, and pain at the thought of Hazrat.

I had no idea if I was driving closer or further away from my brother. I could only hope against hope that he was further ahead and that the next destination would be where I would find him.

The driver took a pair of sunglasses and a magazine from the glove box. "Put these on."

I'd never worn sunglasses before. They felt fun to wear. And I really liked the way I didn't need to squint in the afternoon light.

"Read the magazine and ignore what's going on around you."

"But I can't read Farsi very much," I said, puzzled for a moment by the role the magazine was supposed to play.

"It doesn't matter. It's a car magazine—just look at the pictures and smile at the guards."

"Um, okay," I said, beginning to think it would be easier in the trunk.

"*Salâm aleykom,*" he said, his Farsi accent strong. "Now you say it."

I knew this of course. Every Muslim knows the Arabic greeting, "Peace be with you." Most Muslims around the world use the simple, shortened version, "*Salaam,*" peace, as a way to say hello.

"*Aasalâm alaikum,*" I responded. I tried to say it in the best Farsi pronunciation I could muster, but clearly it came out with a Pashtu accent, with all the emphasis in the wrong places.

He groaned. "Say it like that and we'll get arrested for certain. Try to sound like less of a peasant, and more like you belong both in this car and in Iran."

I bristled at that. "*Asalâm aleykom.*"

"That's better. Not like an Iranian, but better. Please just smile and look at the magazine. Say nothing unless you are forced to."

When the moment came at the checkpoint, I was shaking so much I could barely hold my magazine still. Being in the car was no longer exciting—I just wanted it to be over.

I watched a truck full of migrant workers being searched in front of us. I could recognize them as Afghans. They had papers, which the police studied in detail. The police pushed them around and handled them roughly. I was terrified of what treatment we were surely about to get.

"Remember what I told you, boy," said the driver, through the corner of his mouth.

The border guard approached his side of the car.

"*Salâm aleykom.*"

The steely-faced police officer ignored the driver's greeting. "What's your business?"

"My brother and I are visiting relatives for the day."

"Oh, yes?" said the guard, as if he had his doubts.

I felt the guard's eyes on me. I lowered my magazine.

"*Salaam*," I said, in as bright a tone as I could muster, hoping it came out the Iranian way. I was shaking inside and out, terrified. I tried to control my hands to hold the magazine steady.

"His mother wants me to discuss a wedding match for him, with my cousin. He is getting of an age. That's who we are going to see."

"What's your favorite car, boy?" the guard demanded.

I pointed to a smart black police Land Cruiser parked by the gate.

He grunted with approval. "If only choosing a wife was as easy, hey." The two men laughed.

"Okay," said the guard, sweeping his arm. "You can go."

SHORTLY AFTER GETTING A SAFE DISTANCE AWAY FROM THE police security check, the driver made a quick phone call to whoever was waiting up ahead. We drove on for another fifteen minutes or so before we came across a blue open-backed pickup truck, which was parked by the road, waiting for us. It was the type of vehicle most commonly driven in the Kurdish-inhabited mountainous regions. I assumed this meant we were heading back into Kurdish territory.

"Come on. Jump in," the driver of the new vehicle urged.

I was first in, diving onto the backseat.

Mehran and Jawad were struggling to get out of the trunk of the first car, and were not helped when the driver slammed the trunk shut again as another car came around the corner. From my hiding place I looked on, horrified, convinced he'd just chopped my friends' hands off.

The car passed and he opened the trunk again. "For God's sake, get out of my bloody car."

An angry Mehran snapped back at him, "We were trying to do just that until you shut the lid on my head."

I laughed with relief. If his mouth was working, that meant he was fine.

Mehran and Jawad got in next to me, both bitterly complaining about their bad backs and cramped legs.

"If I was afraid of coffins before, I really am right now," Mehran said, rubbing the small of his back. "That was awful."

I started to snigger.

He threw me a furious look and waved a fist. "I mean it, Gulwali. You laugh at me one more time—"

We were interrupted by the sound of the silver car and the rude young man disappearing into the distance. The sound of the engine still gave me a small thrill, even as the racing car vanished out of sight around a bend.

Ensconced in our new vehicle, we wound slowly upward for the next few hours until the driver of the blue car dropped us off in a steep valley, the splintered hillsides sparsely thatched with coarse thornbushes. A trail of pea-green willow trees marked the passage of a stream that tumbled below us.

"Hey."

A voice drifted up on the cool dusk air.

"Hey. Over here." The driver wordlessly pointed us to get out of the vehicle and go in the direction of the voice. We scrambled down a rocky embankment to a fast-flowing stream; a leather-jacketed man who owned the voice was standing on the banks.

"Come. Follow me."

One of the strangest things about this journey was how whenever a smuggler or driver gave us an instruction, we simply followed it. Whether it was *get in the car, stay silent, follow me, eat this, shave your beard, hand over your passport*—we simply followed orders. Without questioning or really even thinking, we put our lives into the hands of strangers, time and again. We had no choice. When they said come, we little lost sheep had to follow.

It's very hard to explain the feeling of repeatedly putting your complete trust in the hands of strangers who see you as a commodity. Every time I did as one of these men asked, I had an acute awareness that this could be the last instruction I would ever follow. Each of these men had the power to take us to our deaths, at any time.

But I knew Allah was always with me. I prayed often, I talked to God, I found comfort in my faith. I don't even know if that was a choice I'd consciously made. I simply had to. Faith was all I had left.

The man led us a few hundred meters to where, at the fork of the river, under a low-hanging willow tree, stood a tiny crooked shepherd's hut. Its gnarled wooden door hung wide, clinging on by one remaining hinge.

"Wait in there." He pointed at the hut. "I will come soon."

Nervously, we entered the dark interior of the hut. The dirty floor was uneven and scattered with leaves and the remains of cigarette butts, long since gutted for any scraps of tobacco. There were a few puddles of clothing in the corners that had been used as pillows, and ancient evidence of a fire in the crooked fireplace.

It was much colder in this mountainous valley. My T-shirt felt as thin as tissue paper, and I rubbed my shoulders to keep warm.

Mehran scrabbled about on the floor, trying to find a cigarette butt that still contained some tobacco. I rubbed and massaged my shins, knowing that we'd be on the move again before long. We were most definitely still in Iran, but where exactly I couldn't tell.

We spent several hours there, until the day turned to night. We were huddling together for warmth when two men carrying hissing, crackling walkie-talkies burst through the broken door. One of them started to jabber instructions at us in Farsi, but they spoke with the heavy accents we recognized as unmistakably Kurdish. "Soon we go. Gather your things. But quietly. Very quietly. Many soldiers and police."

The second man spoke into his walkie-talkie. The hiss and static made it almost impossible to hear. It sounded as if he was talking in code. "The little birds are in the nest. When do they fly?"

We left the hut and formed into a small column. Again, we followed orders, unquestioning, readying ourselves for the next ordeal.

The pair of smugglers had a huddled conversation before the second man evaporated into the darkness.

"We wait," the remaining man said, smoothing his mustache and lighting a cigarette.

Before long, his radio belched two bursts of static.

"Let's go," he said, grinding his cigarette into the ground. "Say nothing and do as I do. Many police looking for you. No talking, no smoking, no farting. You make noise, you go to prison."

We nodded solemnly.

"Follow me. Where I walk, you walk. When I stop, you stop."

We nodded again.

"Let's go."

He moved a lot faster than his age and build suggested. My anxiety levels skyrocketed: if this guy was hurrying because we were at risk of being caught, then, I reasoned, we must really be in danger. I didn't want to go back to prison—any prison, but especially not an Iranian prison.

We snaked our way along a narrow sheep track that followed the stream. A crescent moon sliced through the cloud, giving enough light for me to see the second smuggler standing in silhouette a few hundred yards ahead, on higher ground. He was scanning the horizon behind and beyond us, issuing instructions to the man with us on the walkie-talkie. It felt like a well-organized military operation.

As we walked, all I could hear was the gentle padding of cautious steps on well-trodden dirt, and the occasional crack of twigs snapping underfoot. If the guide ducked we ducked, if he jogged we jogged too; when he stopped to hide so did we, and when he slowed to a more relaxed pace we breathed a little sigh of relief and did the same. We copied him as if it were a child's game of Follow the Leader. If we hadn't been so petrified it might have been funny.

The guide stopped suddenly, causing me to crash into Mehran's back, almost knocking him over.

"Gulwali. Watch it."

"Quiet, you idiots."

Somewhere ahead in the dark, a walkie-talkie rasped.

"Get down and shut up," the guide whispered.

I fell down hard, knocking my hand painfully on a rock as I went. It was all I could do not to cry out in pain. As I lay in the black grass with just the sound of my own breathing in my ears, I stared up at the moon. I wondered if my mother could see the same moon right now. For a brief second, I wanted to scream at her, to tell her she should have never sent me away.

"Get up. Move it."

The guide seemed to be getting increasingly nervous as his walkie-talkie crackled out new instructions.

We continued in the same way, running and crouching like infantry soldiers entering enemy territory. At last, the guide led us off to the side of the track and under some trees.

I couldn't believe my eyes. Sitting under a tree were four young men, Iranians. I panicked—for a second, I thought they might be secret police. Then I realized they looked as scared as we did, maybe even more so.

Our guide addressed them, "You are Ralph's people, yes?"

They nodded at him in confirmation. "Yes, but who are you?"

"Never mind who I am. You need to get up and follow us."

They looked at each other, hesitating.

"Hurry up. I won't wait for anyone," the guide snapped. "Stay if you want to stay, but the way you need is this way."

They followed us. Our column was now eight people strong: the guide, Jawad, Mehran, me, and now the four young Iranians, who were the people of somebody called Ralph, someone I assumed was an agent like Qubat.

The narrow path we'd been following opened out into a wider, fertile valley. We had just set out across it, making our way directly through the center, when suddenly gunfire rang out across the hills above us.

"Get down. Police. Police."

It was chaos: we got down, but the Iranians ran back to the trees for cover. The shots rang out again. It was impossible to work out if they were shooting directly at us, or across the valley at each other.

"Get up. Fast. Run."

We ran for so long through that valley I thought my chest was going to burst and my kidneys explode. Every single sinew, fiber, and muscle in my body hummed with pain. I was desperate to stop and rest but I knew if I attempted to do so for even a second, the guide would leave me to the mercy of whoever was up there firing indiscriminately. I ran like one of the frightened wild rabbits that burrowed along the edge of the valley path—I had no idea where I was running to, only that I had to get away from the immediate danger. I knew with complete certainty that I didn't want to die, not here, not like this. Running for my life was becoming all I knew.

Even now, I have only half an understanding of the routes that I traveled; my memories are often a blur of faces, landscapes, half-formed thoughts; and then there are some moments that are etched on my mind forever. These are the ones I know I will never forget.

Running for my life along that valley in the dead of night is one of them.

Just when I thought I couldn't run any more and that I was going to fall behind, be shot, and die alone on that path, we came to the end of the valley and onto a track where the same blue four-wheel drive we'd been in that morning was waiting. I could hardly breathe after so much running. My lungs felt as hot and tight as one of the brick kilns that scattered the countryside around my home in Nangarhar.

"Get in the vehicle," a voice ordered.

We all ran, cramming seven bodies in as fast as we could.

CHAPTER 18

As far as I could make out in the moonlight, as I gazed out of the car window, the landscape was very different from the terrain I'd seen when I first arrived in Iran from Afghanistan and was taken to Black Wolf's farm. It was less populated, the buildings were simpler, and it was much more arid. I guessed that we were approaching the Turkish border through a different region this time.

We spent a comfortable night hosted by a family who owned a fruit farm. The owner told us he was Afghan by descent but his family had lived in Iran for generations. His family may have been in Iran for a long time, but they still had the old ways: they made us proper Afghan-style scrambled eggs with onions, served with delicious homemade bread and a salad of cucumber and tomatoes. It was some of the freshest, nicest food I had eaten in weeks.

My stomach was slowly recovering from the meager prison rations, so this time I was able to wolf it all down with gusto. The farmer was warm and polite, treating us like guests—he gave us some delicious oranges from his trees. After all that running I was so very thirsty, so it was a joy to feel the juicy fruit exploding across my tongue.

It was surreal. One second I was being shot at; the next I was eating eggs and oranges.

It still didn't cease to amaze me how we had moments like this, or how many seemingly ordinary people and families along the way were involved in these smuggling operations, offering their homes for shelter or safe passage. I don't know if the farmer was paid to host us or if he was doing it because he was a relative or friend of one of the smugglers—they weren't questions we could ask.

I did discover more about our new Iranian associates. They told us they were students and were fleeing Iran due to the political situation there. They were obviously very scared, made more so by being shot at. They said they would only be able to relax a little bit once they were out of their country and across the border into Turkey. I felt the same way.

We got a decent night's sleep in a comfortable anteroom at the back of the farmer's house, and set off again in the middle of the next afternoon.

We drove up ever-higher winding tracks for the rest of the day. Through the windows I could see people working and watering the irrigated fields that stretched as far as the eye could see. The lengthening shadows told me it was the crossover between late afternoon and early evening, the time of day when the sun graces the earth with a golden good-bye before turning in for the night. Entire families tilled the fields, the women wearing brightly colored head scarves. Children played among the crops as their parents toiled. An elderly man and a little boy herded a flock of sheep across the top of a field. The sight of them made my breath catch in my throat like a stone of grief. In another life, that would have been me and my grandfather.

With a pang of guilt, I thought of my mother. During that last, awful twenty-four hours, running for my life, I had cursed her and questioned again and again why she had forced me to go away. Now I reminded myself that she had done it for me, for my safety. Once again I had survived, God had kept me safe. That had to be for a reason. I couldn't let her down now. This journey *had* to mean something. Besides, my brother Hazrat was still out there somewhere, and I had to find him. There was no going back until I did.

As the sun bowed its golden crown into a gloriously beautiful dusk, we continued up the rough gravel mountain roads until we reached a very small and pretty hamlet, nestled up high, with little stone houses carved into the rocks.

The seven of us were led into a house, where an intoxicating aroma of cooking tantalized our grateful senses. And, after a wonderful meal of roast chicken, rice, and naan—our second great meal in less than twenty-four hours—I began to feel so much better. I had a sense that this family were gentle people, something that became clearer when, after the meal, they began to talk to us. They told us we were in a village above Maku, in the remote mountains surrounding the city. This was news I didn't enjoy hearing—Maku was where I had been imprisoned. I hadn't expected to be going back there, not after that dramatic escape. But, I reasoned that if it was where I had been deported back into Iran from Turkey, then it was also where I could get back across the very same border and into Europe again.

The family was made up of a youngish couple with their two small children, and the man's elderly mother. As the food had been served, I had noticed the old lady giving me glances of concern throughout. She spoke only a thickly accented Kurdish that I couldn't understand, and no Farsi, but her son, who did speak Farsi, translated for me.

"You are so young. Where are you going from here?" she asked.

That question threw me. I'd been running so long I had no idea. I had stopped thinking further ahead than the next minute. "I don't know."

She said something back to her son, who nodded at her, sadly. With that, she gathered up her robes and left the room as her son translated for me: "My mother said you should not be traveling alone. It's not right for a child."

There was nothing I could say to that. She was right, but how could I begin to explain? The man told us his name was Serbest, laughing when he said it meant "royalty" in Kurdish. Serbest told us his mother didn't want him to do this type of work, explaining that it was very dangerous for him. Not only did he risk arrest from the authorities for sheltering illegals, he lived in fear of the powerful regional agents and the various local smugglers and drivers who worked for them. While it was the local-level smugglers who brought their charges to him, he knew they worked for the more powerful, wealthy people—people who could easily do him harm if they so wished.

He said the smugglers often tried to cheat him—paying for two guests but instead bringing ten. He said in those circumstances he was left with no choice but to feed the extra mouths. His mother would not allow it any other way.

Serbest's honesty and vulnerability made us all like him. We too knew what it was like to be tricked by the smiling liars in leather jackets and Land Cruisers. I thought back to our first ever Turkish agent, Malik, the liar in the smart suit who was also involved in trafficking women across borders to work in brothels. I recalled the awful chicken coop he'd kept us locked up in. I thought too of the family living there—the kind old lady who had been so nice to me, and her useless, drunken son. Both that old lady and her son, and Serbest and his family, were making their living from harboring desperate refugees, but they were different, somehow, from the agents we had met. They certainly weren't getting rich from it. They were poor people who needed work and money. Were they really so different from us?

Serbest explained there was no other option for him. During harvest time there was some work on the farms, but it was occasional and didn't pay enough for him to save the money to see his family through the long, cold winters they endured. If a family didn't have enough wood, grain, and rice stored, they would go hungry and cold.

It made me realize that life in the remote Kurdish-inhabited parts of Iran or Turkey was no different to that of rural Afghanistan. Yes, there were some bad people—criminals and kidnappers, but most people were decent. Living the same hand-to-mouth existence, they put family, morality, and duty first. And through that they survived.

We all said evening prayers together, then Serbest told us to get some rest because we'd be leaving within the hour.

I groaned inwardly—"*Please, not another journey into the dark.*" Too nervous to rest properly, I instead reflected on the people I had met over the past few months and all the things I had seen—the brutality, the injustice, the poverty, the kindnesses, the mixed objectives that most people had. The fundamentalist little village boy from rural Afghanistan had seen and heard so much that it was impossible to have remained the same.

Meeting people like Serbest had helped to open my mind. As had so many other things on this journey, from praying with Shia Muslims, to hearing stories of poverty in Africa, living in modern cities, and seeing how other cultures lived . . . The boy who used to boss his aunts around for fun was long gone.

At that moment, for some inexplicable reason, my maternal grandfather's face came before me—the grandfather who was an imam. I heard his voice in my head so loudly and clearly it was as if he were standing right next to me. It was unsettling but not frightening. I swear I heard him say to me, "Life is an education, Gulwali. And all life must have a purpose."

WHEN THE OLD LADY CAME TO WAKE US, THE WIND WAS whistling through the trees outside. She and Serbest's wife held a Quran above our heads, which we walked underneath as they said prayers of supplication, asking for God to keep us safe, and for our return. My grandmother had said similar prayers, a tradition for travelers, before Hazrat and I had left our house for Waziristan and the start of our journey—when I had thought my journey was just for a holiday and that I would be home soon. The memory was heartbreaking but I was so touched that these kind people bothered to pray for strangers like us.

The old lady placed a work-coarsened, wizened hand to my cheek and stared at me with tears in her rheumy eyes.

I heard the sound of hooves and whinnying.

Serbest grinned. "That will be your gift." He slipped into the darkness. A few minutes later, he came back riding a gray horse, holding on to the reins of a second, sturdy-looking brown horse behind his. Both horses wore an embroidered bridle that brought back memories of my grandparents' *kochi* "gypsy" tent. He smiled at me. "It's for you. You are too young—you will not be able to walk this path."

"No way."

"It will help. Trust me. But I am sorry, I cannot pay for him. He will cost you twenty-five dollars."

"No. Forget the money. Forget the horse."

I still had my dollars so I could pay for it, but in no way did I want to: I had been around donkeys and horses as a boy, but I had never really liked them. They scared me.

Everyone was laughing at me and my face burned red with angry humiliation as I backed away. "No. No. *No.*"

Mehran just stood smiling, trying to hold back his laughter.

Serbest held out the reins again. "Come on, I will help you up."

AN HOUR LATER I WAS SITTING ON THE HORSE, FILLED WITH oceans of gratitude both for Serbest and the animal.

The journey was brutal.

From the village we set off on our way, crisscrossing the landscape. The irrigated valleys and fields below had given way to sharp rocks and steep little donkey tracks that were barely passable by foot. The surefooted hooves of my mount were far steadier than my boots, my so-called best friends, could ever have been.

"Get off for a bit, Gulwali. I need a turn."

I turned to Mehran. "Sure."

It was only a couple of hours' walk before we reached what Serbest said was the meeting point.

It was like an ancient battle scene or, rather, a battle-preparation scene. Literally hundreds of people, men and women, were gathered in a clearing. Some were resting under trees, bundled clothing under their heads; others were preparing themselves—lacing up boots and putting on warm clothes.

It must have been close to midnight, but the moon was bright and I could tell that all of these people wore the same tired and confused expressions as I did. They were refugees. I tried to work out where they might be from. No Africans this time, but lots of Arabs and Asians—Afghans, Pakistanis, Iraqis, and what looked like more Iranians. Their local guides, whoever their Serbests were, stood beside them, watering their horses, preparing saddles and packs for the journey ahead. I seemed to be the only refugee on horseback.

We waited there for over an hour, giving me time to absorb the whole scene. More and more people started to arrive on foot, a few on donkeys. The clearing was totally inaccessible by vehicle.

What I didn't understand was how so many people could be there without anyone seeing us; there was so much noise I was sure the whole city of Maku, somewhere below us, would hear us. I was so scared the horsewhip-carrying police, or worse, the kidnappers, would come any second.

I also thought about the first time I crossed from Kurdish Iran into Turkey. Then, we had walked across the border on an ancient trading route. This time there were no jovial, singing merchants trading their wares. I recalled Black Wolf's nephew theatrically banging on his drum as he sat astride his horse like a tenth-century warlord. This time, however, the mood matched the arid landscape. It felt somber, almost doom-laden.

I don't know who gave the orders for when to set off, but it seemed that someone was in control because when we moved, everyone started walking together—a long line of people, mules, and horses winding up the trails, across roads and rivers. We were quite close to the front of the group and, as I looked behind me, all I could make out was a long trail of exhausted people.

Serbest was a good and generous guide. He rode his own horse close to us, and kept checking to see if I was okay.

"Good boy. Thank you, boy." I patted my horse. Without him, I am not sure I could have made it. At times the passes were so steep that people had to hold hands to avoid falling down onto the rocks below. Mehran, Jawad, and I shared the horse; whoever wasn't riding him held on to his tail for safety.

The journey went on for most of the night as we wound around tracks that took us higher and higher, to where the air got so cold it was hard to breathe. Unbelievably, at some point in the night, we crossed over two busy roads, cars coming from both directions. How could these hundreds of the walking hopeless and hungry not be detected?

A frightened voice suddenly rang out in the darkness: "Get back. Back. Everyone move back."

There was no back, only back down the narrow track—the way we'd come. All I could see behind me was more people. Confusion reigned. It was pitch-black and no one could see anything as they tried to follow the order, scrambling to turn around. Miraculously, the horse found its way to the side, behind some rocks, where we both sheltered.

As quickly as the order had come to move back, a new order came to continue moving. I am absolutely sure some people got left behind in the confusion.

We had walked steadily for another thirty minutes or so when a shot rang out.

"Take cover. Cover."

Most people managed to hide behind boulders. I was thankful the horse stayed calm—I suspected it wasn't the first time my trusty mount had heard gunfire. As we tried to shelter, some people said the shots were coming from quite far away, from the other side of the cliffs, on the Iranian side.

I couldn't understand that. Our army of the desperate was clearly walking away from Iran into Turkey, so why shoot at people who were leaving?

In the chaos, we lost Serbest. I panicked, trying to control the reins so I could find him. "Serbest. Where are you? Help me."

When his horse caught up with us, his face was ashen with worry. "Little one, thank God you are safe—my mother would never have forgiven me. Keep your horse behind me and don't fall back."

As the sun awoke to a new day and peeped through the veil of darkness to bring dawn, I saw that we were almost at the top of a mountain range.

"Here is Turkey, here is Iran." Serbest gestured first to the east, then toward the west. All I could see was mountains and more mountains.

He told us we were only a few miles from Turkey now. Europe was once more in reach.

Serbest pulled the reins of my horse to a stop. He dismounted from his own and began shaking hands with his seven charges one by one. "This is where I say good-bye. Farewell, my brothers. God will be with you. Stay on top of the game." He turned to me. "And especially you. Go well, little boy. Go well."

I was sad to say good-bye to our new friend; I had liked Serbest very much indeed. But if I was sad to say good-bye to him, then I was surprised to learn that I felt even more sad to say farewell to my horse.

That horse, the simple creature that I had been angry at being forced to ride, had been my savior that long night. He was so gentle with me, and so sure-footed, and sitting astride his back had been the closest thing to safe and secure that I had felt in a long time. He had reminded me of the lovely chestnut mare in Black Wolf's compound. That brief interlude of calm and comfort when I had breakfasted with his family seemed so long ago and far away now. Now I couldn't bear to see the horse go. I was annoyed to realize I was fighting back tears. "Don't be a baby, Gulwali. It's just an animal," I chastised myself.

We stood and watched as the various agents, smugglers, and brigands walked back to Iran. Serbest stopped and turned to wave us a cheery good-bye. Then he and the horse were gone from view.

Those of us left behind paused to rest a little longer before walking forward into Turkey. But I still couldn't understand just how this many people were supposed to walk right into Turkey undetected.

Spread out before me was a very different view to the one I had seen the first time I had stood in the no-man's land between Turkey and Iran. Then, just a couple of weeks into my journey and still a frightened child, I had been transfixed as I had looked to one side of the horizon and seen night, on the other side day. This time the rocky landscape looked the same, and it was impossible to tell how and where Turkey differed from Iran.

But I felt a new sensation, as though I was standing between two worlds—the old and the new.

My old and *my* new.

I calculated it must be close to seven months since I had left my home and my family. I was still no closer to finding Hazrat and unsure as to where my journey ended. But I was certainly changed, no longer a little boy. As I stood there, with Iran to one side of me and Turkey, Europe, just a short walk ahead on the other, I told myself that this was not a place I wanted to be in for the third time.

"This time, Gulwali, you will make it. You *will*."

CHAPTER 19

IT WAS ONLY WHEN THE IRANIAN GUIDES AND SMUGGLERS left that I had a chance to really see who else was with us.

There were a couple of hundred other people: most of them huddled in small groups, just like Jawad, Mehran, me, and the Iranians were. Their ages spanned the range of young to old. Their faces showed they were from many different places. By now I had a good understanding that people were fleeing for a variety of different reasons—some because of conflict, others because of poverty.

Many carried small shoulder bags and cases, no doubt filled with a lifetime's possessions. Others had nothing but the clothes on their backs—as if they'd left their houses that morning, and were now stuck in Turkey. Some had sturdy boots like mine, which they peeled off, allowing their wet, blistered feet to breathe. Others had plastic sandals—how they managed to walk any distance in those, I don't understand.

Most of the men were bearded. Shaving was an infrequent activity, and a low priority for most.

I was dismayed to see a couple of women with small children. The children were dirty, with runny noses and matted hair. One little girl wore a blue bobble hat, but only had a pair of leather sandals on her tiny frozen feet. My heart lurched when I saw that. I wanted to go talk to her, to say something comforting. But the way her mother held her so close, so fiercely, stopped me. She watched the men around her with cold eyes. Like a lioness protesting her cubs, she just stared, looking for any signs of danger. I decided it was better to let the family have some privacy.

What quickly became very clear was that not a single person among us had any idea what would happen next. The Iranian guides had simply pointed down the mountainside and told us to continue the walk to Turkey.

"Is this familiar?" Jawad looked at me, as if I should know where we were.

"How should I know? It looks nothing like the last time I was in Turkey."

"How do we find the agents?" he persisted.

"Why do you keep asking me? Do I look like I know?" I know he wanted answers, but he was getting on my nerves.

A couple of groups stood up and started to walk away. In an information vacuum, even the appearance of knowledge has power. I don't really recall anyone else saying much, but the ragtag herd of human beings got in line and followed.

After a short distance, we could see men in brown uniforms approaching us on horses. They were armed.

I began to panic. Their uniforms didn't look like any I'd seen before—none of their clothing matched properly.

Mehran and I were right at the front of the column. We started to hold back, getting ready to run into the trees if things turned bad.

The riders waved and gestured us to come close.

I recoiled: mafia kidnappers—I was certain of it.

They didn't introduce themselves. They didn't say a word—not even to check how many of us there were, or if we were safe. Neither did they wait to see if we followed—instead, they just turned their horses around and started down the hill.

I wanted to run. How could we trust them?

Jawad looked at Mehran and me, following our lead.

"Stop," I said. "They are kidnappers. Let's run back."

He paused. "I don't think so. This is all too organized."

"Don't risk it. Let's go back. Come on." I tugged at his hand, trying to pull him away.

"Gulwali, I think it's okay. If they were kidnappers, there would be more of them."

"It's not okay. They will kill us. Do you want them to cut off your ears? Come on, let's go."

As I made to leave, Mehran yanked my arm back so that I was standing in front of him. He was always quick to anger: "Go back where, Gulwali?" he shouted, his face flushed.

He had a point.

"Do you want to go back to Iran? Walk back over the mountain? You have no horse now. We've just walked for a whole night, from only God knows where. So do you suggest we go back to God knows where?"

"I know. But if they kidnap us—"

I didn't like the fact that I was being held responsible for this decision. The truth was, I had no idea whether it was a good idea to follow them or not.

As the other men streamed slowly past us, reluctantly I fell into line behind them. Like lambs to the slaughter, we carried on down the mountain.

For the next couple of hours, two hundred exhausted people walked behind the men in the homemade, fake-looking uniforms. The night had been so cold, but by now the harsh mountain light was beginning to sting my eyes. I was worried that the authorities would see so many people in broad daylight and arrest us, but the men leading didn't seem bothered at all.

We continued to walk, until we reached a handful of filthy cattle trucks. The waiting drivers herded us up the ramps and into the back. They pushed and shoved until the truck was so crammed there was only room to stand. It was impossible to breathe, and my ribs ached we were packed so tight. There was a huge diversity of nationalities among our number—the hundreds of people represented the collective enterprise of different agents: each one had their own little flock. As far as I knew, Qubat, back in Kabul, was still our main agent. The middlemen identified us either by Qubat's name or the name of his main representative in a country, such as Malik. One thing was clear, this was a highly organized infrastructure. Despite the massive numbers of people on the move and the vast distances we traveled, the system was pretty efficient. I felt constantly vulnerable but where I felt most unsafe was in the hands of the different drivers, most of whom I guessed had been recruited locally and didn't know who they were working for, or at the handover points when we had no idea who we would be passed on to next.

THE JOURNEY ON THE TRUCKS QUICKLY BECAME UNBEARABLE. We'd had no food or clean water during the night—the only time we'd drunk was from mountain streams. I hadn't eaten properly since our final meal at Serbest's house in Iran, less than twenty-four hours ago.

Our convoy wove through mountain passes and valleys. After a couple of hours, we pulled over and were ordered to get down. Many of the passengers were now so weak they could barely stand, let alone jump out of the stinking truck.

The Afghans, who formed the majority of the crowd, were separated from the rest of the group by the drivers, while the Kurdish, Iranians, Iraqis, and Pakistanis, along with the couple of young families, were immediately put back into one of the trucks and driven off. I caught only the briefest glimpse of the little girl in the blue hat. Her grubby face was wet with tears and my heart broke for her.

The other two trucks drove away empty.

A little further along the road we could see twenty or so Afghan men already there, sitting cross-legged or lying on the ground.

One of the refugees shouted in our direction. He spoke Pashtu.

"Brothers. Bad luck that they dropped you here. We've been here a few days already. It's bad luck."

The group looked depressed and particularly hungry.

We sat down next to them.

They were as confused as we were about where they were. They told us they had been there for days, sitting outside during the day without food or water. At night, an old man had come on horseback and taken them down the valley to his farm. There, in a stable piled high with dung, they had slept on tattered blankets and had been given only a tiny portion of what they described as "food so disgusting, just to look at it makes you sick."

I felt sorry for them. But I was also irritated. We hadn't exactly been having it easy ourselves, and they seemed to think we should know what was going on. "Who is in charge? How can we get out of here?"

I snapped. "We have just walked over the mountains. How do we know?"

I went to the stream and washed my feet in ablution. I needed to pray. I didn't know how or when I would be on the move again, but I think blind faith helped to drive me on. I had a growing sense of fate, of a belief in God's plan for me. The water was icy cold and clean. I drank my fill, then washed. It made me feel better.

That's when I knew what I needed to do. I walked to the group, addressing them as calmly as I could, "Brothers, who will lead us in prayer?"

No one offered. The first group looked at me blankly. Why was a kid asking this?

"Jawad, please can you?"

He nodded, stood up, and began to lead us in *jammat,* or congregational prayer. Usually for this type of prayer you have a mat for prostration—for putting your forehead to the floor. Instead, we used our clothes or bags. We supplicated to God with our palms outstretched together.

After the prayer, I like to think some of their hopelessness dissipated. We sat in a circle working out a plan. There was a lot of argument about what to do. It almost felt like a *jirga*—the traditional gathering of elders and respected men to resolve disputes and make judgments. In happier days I had observed many *jirgas* held in our house, listening to the men talk and debate unresolved community or political issues. Whenever there is a group of Afghans you can be sure there will be zero agreement. Arguing with each other is just something we can't help. I think it's in our blood.

"We have to get out of here as soon as we can," said one of the first group.

"And how are we supposed to do that?" asked Mehran, rolling his eyes.

"Soon the old man will come. He will ask you for money because he always does. Trust me, you don't want to argue with him. You will be beaten."

His friend nodded in agreement. "His men beat us when we disagree. Staying there at night is the worst thing that can happen to you."

"Just pay him the money and don't argue with him. Whatever happens, you need to pay him."

Mehran, Jawad, and I were beginning to understand why these men looked so miserable. This was not a situation we wanted to be in any more than they did. But, because I was calm following our prayers, a sense of purpose came over me. Just as in the prison yard in Maku, when I had felt it my duty to negotiate our way out and take the people without money with me, I believed the three of us—Mehran, Jawad, and I—had come here for a reason: to rescue these men and take them with us.

On this journey everyone was out for themselves; in part that was because they were so helpless and powerless. No one ever believed they had the power to make change happen or influence the smugglers. I felt that by acting as a group we could make it through somehow. If we showed the old man a united front instead of everyone shouting, crying, and pleading, we might get out of there. I wanted the old man to know we had nothing left to lose. And I wanted to look him in the eye and make him understand.

I was dozing when I heard the sound of clicking hooves on the road. A gray horse with a white-haired man sitting astride it. He looked old but strong, like a seasoned warrior. By his side were three or four younger men, also on sturdy horses.

The old man dismounted. From his saddle bag he took out a few pieces of bread and a couple of raw onions. He turned, about to throw the food in our direction. He wanted us to jump on it like animals.

"*Salaam,* Uncle," I said. In my culture "uncle" is a standard term of respect for an elder male.

He looked at me, surprised.

"Please give me the food," I said. "I will distribute it."

He spoke back to me in a mixture of Kurdish and Farsi. "You are the new people. You don't seem to know the rules."

I glanced around; the others were all staring at the ground. I could feel the old man tensing with anger. I was truly scared, but I carried on. "Apologies, Uncle. I only sought to help you."

He frowned, then passed me the bread.

I gave it to Mehran and Jawad, who started tearing off pieces and handing it around.

The old man continued to glare in my direction, before his voice erupted like thunder. "You new people. You need to give me money for the food and shelter."

At this Mehran snapped. "What shelter? And you call this food?"

The old man stared at Mehran, but let the slight pass for the moment. "Five dollars each, man. No pay, no protection."

Despite the warnings from the first group to do as he asked, most of our group started yelling at once, as several voices rose up in uproar.

"You've already been paid by our agent."

"Why did you drop us here?"

"What kind of trick is this?"

"You are not getting a penny."

The old man looked completely unfazed. His men stood by, tense like wild dogs.

"No pay, no protection," he repeated, a calm menace in his voice.

Jawad reacted smartly to deflect the situation. "Okay, Uncle," he said. "Let me collect the money for you."

The three of us walked around taking money from all the new arrivals.

Some looked at us with a sense of betrayal. No one wanted to pay, but something inside me told me I was doing the right thing and I could get us all through this. I knew we weren't going to get out of here by arguing, but get out of this situation we needed to do. I felt only a determined stillness as I let my instinct take over. Jawad and Mehran didn't really know what was going on in my mind, but I got the impression they trusted my judgment.

The white-haired old man stood watching us quietly, almost amused by the scene.

Jawad and I walked over to him with the bundle of cash. "Here you are, Uncle." We smiled and, in our best Farsi, thanked him for giving us his protection and for bringing us the food.

The old man eyed us suspiciously. Such obsequious behavior was clearly not the norm.

I took a deep breath and prayed my plan was going to work. "Uncle, with respect, we would like to leave this place. Is it possible for you to arrange transport for us?"

For a few seconds he stared straight at me, then turned to his men. I was expecting the blows to come raining down at any second, but instead he burst out laughing. "Well, that is something. In all these days, I never saw the animals behave so calmly or obediently." He looked back at me. "I don't know how you did that, child."

He looked back at his men. They all laughed too because he did. I perceived that everyone, even his own people, were terrified of this old man.

"As it happens," he continued, "I had planned to send you all tomorrow anyway. But seeing as the boy asked so nicely, I will allow you to move today." And with that, he and his men got back on their horses and cantered away.

As soon as they were out of earshot I turned to the others, expecting a round of applause.

"You stupid little boy. Now he's got our money and he won't come back."

"You offended him. Now he'll sell us on to who knows what."

"Are you so foolish to believe his lies?"

I sat down, completely despondent. I'd done my best to help. Couldn't they see that?

As the hours wore on, the abuse continued.

"So, where is your truck, Gulwali?"

"I bet you made him so mad, we won't even be allowed into his stable tonight."

It was getting to me. I shoved my face into my arms so no one could see my tears. I didn't know why I had acted that way with the old man and, in hindsight, maybe I had been naive. But, in the moment, it had felt so much like the right thing to do. But I should have known better. Hope, along these roads at least, was an allusion. Something dangerous—and costly—to entertain.

The sun was setting when the sound of engines rumbled over the horizon.

Trucks!

As they approached us, a few of the others stood up and cheered. I stayed crouched, quiet and calm.

With a roar of crunching pebbles, two cargo trucks pulled in before us.

CHAPTER 20

WE CLAMBERED ONTO THE BIG TRUCKS AND HEADED OFF into the mountains.

I admit to feeling a little bit full of myself, but I was trying my best to stay humble.

Mehran was buzzing with excitement at what we'd managed to do. He slapped me on the back. "We are learning this game fast, brother."

I shrugged. I knew that I hadn't acted on any grand plan; I had just done what felt right. And I knew I hadn't wanted to stay in the stable the others had described. It had sounded too much like the awful chicken coop back in Van. To be back on the move again so soon felt like a victory of my willpower over the smugglers.

But Mehran was right about one thing—we were learning. This time we knew where to sit on the truck—right at the back, near the driver's seat. When the driver threw food and water into the back, we'd be the first to catch it. Often there wasn't enough for everyone and if you didn't grab it fast enough you went without. That said, even in those circumstances people did usually behave decently and share, but if the resources were so limited it was impossible.

Mehran started regaling some of the others with tales of our daring prison break: "The police were everywhere. But we just kept running. Yeah, we were scared but, you know, sometimes you just have to take the risk like a man."

I looked affectionately at my friend. He was exaggerating, completely failing to mention how utterly terrified we'd really been. But after the events of the past few days, he deserved this moment of distraction. We all did.

In truth, though, I couldn't relax. Yes, we were moving. Yes, the old man had brought the trucks as promised . . . but it's hard to describe the sense of confinement and worry as you are loaded onto a vehicle, locked in by a stranger, hungry, thirsty, and with no idea where you were being taken, or for how long you might be confined.

I also knew that if anything went wrong on this journey, I'd be the one they'd blame. Some of this group were so angry at their fate I feared I'd be ripped apart at the slightest setback. I also prayed I'd finally make it out of Turkey safe and sound. Having been on the road for nearly seven months, and still only on the very edge of Europe, I could only reflect how the very first smuggler, the man who had taken me to Peshawar, had promised my mother it would only be a few weeks. I now knew it could be months, if not years.

That's if I made it at all.

FOR THE NEXT THREE OR FOUR DAYS, THAT TRUCK BECAME home. It was where we slept, ate, and talked. The days were hot so it was a relief that on quiet country roads the driver was relaxed about having the tarpaulin roof down, which meant we could breathe properly and watch the countryside roll past. Only on the busier roads did he order us to pull it up over our heads so we couldn't be seen.

As we proceeded, I for once had a vague idea what was happening and where we were going: I was increasingly certain we were heading to the city of Van—the city near the border I had been taken to on my first trip to Turkey. That feeling intensified when we again fell back into the familiar routine of walking at night and driving by day. I had played this game the first time I had arrived in Turkey—it was all about avoiding the military or police checkpoints. During the day we drove as far as we could without hitting a checkpoint. As one came close, the truck stopped and, if we were unlucky, we had to stay still and quiet until darkness set in. At other times men on foot, presumably working with the driver, took us to dilapidated farm buildings to wait out the daylight hours. That was much more preferable because we could use the toilet and rest. Once it was safely dark, we would set off walking into the hills and take lengthy semi-circular routes to bypass the checkpoints. Once safely past them, and while it was still dark, we'd walk back down to lower ground where the truck was waiting for us, and we could set off by road again. On one exhausting night we did this stop and start, walking and driving routine three different times.

Though the tactics were the same, the countryside was different. The terrain was rockier, with slippery slopes and much longer—seemingly endless—paths.

I think if it had been the first time I had done this I would have really struggled, but because I knew what to expect I was able to control my mind to push my body through. And, just like Baryalai had done before, Jawad was really helpful to me, holding my hand and pulling me along when I got too tired to put one foot in front of the other.

The only really risky moment came just outside Van, where I had indeed correctly guessed we were headed. A few kilometers outside of the city, the driver shouted back to us to get as low down as we could and pull the tarpaulin up and over our heads. It was early morning rush hour.

Hidden under the thick heavy tarp, we felt sweaty, claustrophobic, and nervous. I couldn't see a thing but I could hear the traffic noise, so knew we were on very busy roads.

When the truck suddenly stopped at what felt like a roundabout, I couldn't believe it.

"This is madness," I said. "What is he doing?"

"Maybe there's a problem," Jawad volunteered.

"Gulwali, why don't you try to look?"

"No way. I am not going back to prison."

"Get ready to run again," said Mehran.

I couldn't believe this was happening. We heard the tarpaulin rustle and the driver's face appeared. He looked white, as scared as we were. "Stay quiet. We've broken down."

In whispers, the message passed down the truck. Everyone held their breath.

We could hear police arriving and talking to the driver. A couple of our group spoke Turkish, but from under the heavy tarpaulin the outside voices were too muffled to understand anything. We all knew that if the police wanted to search underneath it we'd be rumbled.

To our amazement, they didn't even come around to the back of the truck. It was a miracle that we weren't caught.

Somehow we got moving again and were on our way.

By the third day, living inside the truck was becoming unbearable. After our near miss, the driver insisted on keeping the tarpaulin up all of the time. It was suffocating, and the sickly sweet scent of sweat and urine permeated the little air we shared. The only time we could go to the toilet was at night, when we were walking, but even then we had to do it quickly to ensure we didn't get lost or left behind.

For the final twenty-four hours, we didn't walk at all: the driver didn't stop once. I was so thirsty, but I couldn't drink anything because my bladder was bursting. It was pure agony.

JUST WHEN I FELT I COULDN'T TAKE IT ANY LONGER, WE reached our destination. When we arrived at a small suburb on the outermost outskirts of another city, which I again correctly guessed was Istanbul, we were taken to a huge sprawling complex of industrial buildings, situated by a noisy main road. Whatever industry had once happened in these buildings had long been replaced by a new trade in humanity, our bodies replacing the goods kept inside the storerooms. About eighty men, women, and children were already inside, sitting on a damp, filthy carpet floor, all with exhausted eyes. Most of them looked Iraqi. At that time—spring 2007—it was the height of the conflict, following the Iraq invasion in 2003. Hundreds of thousands of Iraqi refugee families were on the move, seeking safety.

It was the second time in a few days I had encountered children. Seeing children was rare, because usually the agents and the smugglers who worked for them kept them separate from the men. Although still a child myself, I didn't count as one in the smugglers' eyes, I suppose because I wasn't traveling with a family. I wasn't much older than some of the kids there, but, after the horrors of the journey and in particular the three stints in prison in Iran, Bulgaria, and Turkey, I no longer saw myself the same as them. I didn't feel like a child anymore. Yet I was so sad for these other kids and their lost innocence.

I stank. And I was soaked with sweat. I hadn't washed since leaving Iran. That was almost a week ago: one night and day to cross the border, a day on the roadside with the old man just after entering Turkey, and then three days in the truck. In that time I had climbed mountains, fallen over in mud, ridden a horse for several hours, and slept next to a hundred other unwashed bodies. All in the same set of clothes. No wonder I smelled so bad that even I gagged when I caught a whiff of myself.

For Muslims, being dirty is a great shame. The reason we take ablution before our five daily prayers is to stand clean before our Creator. Not being able to wash myself was a great source of distress for me, as I'm sure it was for all of the other human cattle kept in that vast, damp room.

At that moment, I felt less than human.

I wanted to run from that place as soon as I could. I did consider it, because by then I had a good idea of how to find my way around Istanbul, so I figured I'd be able to cope. I think I would have persuaded

Mehran and Jawad to try to run with me, but the doors were blocked by burly guards carrying guns. All I could do was sit down and wait.

If my life wasn't about running, then it was about waiting.

My stomach was still in agony. It was distended, swollen, and very painful. Though it had been days since I had gone to the bathroom, I still couldn't pee. An old man—an Iraqi Kurd—suggested that I eat yogurt, as it might help. He offered to ask the guards, who were clearly Turkish Kurds, to bring me some.

I watched as he approached them. "Excuse me. The boy is sick. He is in a lot of pain. Can you please bring him some yogurt?"

"This is not a shop."

"Please. He needs help. He's a child."

"Sit down, old man. You'll all be gone soon."

There had been about 130 men across both our trucks; with the 80 already there in the storeroom, we numbered over 200 again. We weren't given any food, but there were three toilets and one shower—between all of us. With so many people attempting to use the toilets they were overflowing with filth. I wanted to keep trying to make my bladder work, but I really couldn't face going in there. Besides, the queue was massive, and I wasn't sure anyone would have let a lone boy go to the front.

Little by little, drivers representing the different agents came to collect their charges. They identified people by asking the same old question—who are the people of so-and-so? I think I heard some eight or nine different agents' names being used. After a few hours, nearly everyone had been picked up, including all the families and the other men from our truck. By midnight only Mehran, Jawad, and I were left.

The guards were eager to get rid of us. The guard in charge spoke to us in Turkish, while one of his men translated for us: "Your responsible person has been in contact. He says he cannot come tonight. You will have to stay here."

I was in so much pain that I couldn't bear the thought of sleeping on that hard, cold floor. I badly needed water and food. Whatever the next location was, I had to hope it was better than here.

I don't know what possessed me but I stood up and faced him. I shook my head firmly and made a telephone gesture to my ear, little finger and thumb extended. "Call him," I told the guard. "I want to speak to him."

Remarkably he did. He came back with a mobile phone and passed it to me.

The guy on the other end of the phone was Afghan.

"Are you the people of Qubat?" he asked. I could tell from his voice that he was drunk.

"Yes," I replied. "And may I ask who are you?"

"Are you four?"

The question perplexed me. Then I remembered Tamim.

"No, three. One was left in Iran."

"I have no money to come and pick you up tonight. It is far, I need a taxi."

"What are you saying? We need to leave here. Just give us the address and we will come to you ourselves."

"Qubat hasn't paid me."

"That's your problem, not ours. Our families paid him. We are getting out of here tonight. Tell me where to come."

He slurred drunkenly. "Be patient. I will come for you tomorrow. Let me talk to Qubat."

"No, I don't care what you do. But I want to come there tonight."

"But it's the middle of the night," he slurred. "Go to sleep."

Eventually, he promised to send a taxi to pick us up, if we agreed to pay for it ourselves.

To do so, we used some of the $100 we'd been given in Tehran. The driver kindly exchanged what was left of our change into Turkish lira, which we also spent at an all-night cafe to get some food. Finally, I was able to go to the toilet, but the pain in my stomach hadn't gone away.

I had made the right call in demanding that we move. I was relieved to find the drunken agent's *musafir khanna* (guesthouse) was a clean and tidy two-bedroom fourth-floor apartment in yet another sprawling Istanbul suburb. As we entered, we woke up the other refugees staying there, who weren't very happy with us.

The next morning we chatted to them properly—they were five young Afghans.

The agent, his breath reeking of cigarettes and marijuana, arrived not long after we'd all woken. "I'm Amiri. You are the impatient ones. You were very rude last night."

He told us to keep a low profile and not to make noise, and be alert at all times. He gave us permission to go outside, but only if we really needed to.

"Better to stay indoors for your own safety."

The good thing about that apartment was that it wasn't overcrowded. We were able to cook for ourselves and eat reasonably well. Jawad was a great cook and we used the rest of the shared money to buy simple but tasty supplies.

Amiri, however, was a nightmare. He would visit daily, complaining about Qubat and threatening us: "He still hasn't paid me. I am not letting you leave here until I have my money."

Amiri frightened all of us, but I found him particularly unsettling. Whenever he complained, he looked at me directly, as if it was somehow my fault.

"So what can we do?" Jawad tried to reason with him. "We don't even know who this guy is. It's between you and him."

Eventually, Amiri decided to move us all to a different location because it was a cheaper option. I was worried, but I can't say I was surprised. This was my second time in Istanbul, and I remembered clearly the system of smuggling here. It involved a lot of threats and a lot of movement between locations.

Amiri took us to a crowded basement apartment in a rough-looking neighborhood, where about fifteen other men huddled together, waiting for their next move. They were all from northern Afghanistan, and native Dari speakers. They were angry that their already overcrowded space now had to fit in eight more: me, Jawad, Mehran, and the five young Afghans, who were also Pashtu speakers like us. The first group was doubly resentful that we had permission to go out to the shops and that Jawad had been given both a key and a mobile phone by Amiri. The phone had only a tiny amount of credit on it and he'd made clear it was only to be used for staying in touch with him in case anything went wrong. The others had a different agent, a friend of Amiri's, whose name we weren't told. But it seems this man had kept them locked in. We were instructed by Amiri not to challenge the other agent's rules and not to allow them out with us. I can see that must have been very frustrating for them, but we were as scared of Amiri as they were of their agent.

The anger spilled over in disputes about rivalry and ethnicity.

"You Pashtuns represent the Taliban," they yelled. "It's because of your people that we had to leave Afghanistan."

The Pashtun are the traditional rulers of Afghanistan and it is true the Taliban were predominantly Pashtuns. The leaders of the Northern Alliance, a group of non-Pashtun tribes in the north, resisted the Taliban, helping the United States overthrow them.

These men saw the little apartment as a continuation of the war: "Why did you leave your country? Life was perfect for you people."

As far as I was concerned, all Afghans had suffered from the long years of civil war and subsequent U.S. invasion. It didn't matter which ethnic group they belonged to.

The apartment got so hot and with it the tension. It was like a pressure cooker waiting to explode.

I tried my best to keep the peace: "Look, we're just like you. We've all left our homes. And we're all Afghans. Do you think I left for no reason?"

"Forget it, Gulwali," said Jawad. "The Northern Alliance are the ones in control now. And they know it."

The fights raged for hours. Everyone would be yelling and insulting each other. They refused to share their food with us, so we cooked for ourselves on a gas cylinder. It was ridiculous—with both groups cooking separately, the apartment got even hotter.

As the atmosphere in the apartment became more and more toxic, we couldn't handle it anymore. Their agent was a friend of our agent and surely they could sort it out. In the end, Jawad called Amiri. Both he and their agent, unsurprisingly himself an Afghan, came to try to resolve the situation. They weren't happy.

They told us to put up and shut up. "This is not Afghanistan. I don't want to hear this nonsense again," growled Amiri.

Amiri, who was Hazara, was completely right. Here we were, all running for our lives, yet still arguing over the language and ethnic divisions that had destroyed our country and forced us to leave in the first place.

For all of his apparent wisdom, however, Amiri couldn't be trusted. Whenever our group went to the shops, we plotted how to get away.

"This bastard is a joke," Jawad complained. "Not only do we have to deal with these Northerners, I don't believe he's going to take us to Greece."

Mehran agreed. "If he couldn't even pay for a taxi the first night, how can we expect anything from him?"

"But he won't let us leave," I said. "We've cost him money."

We all knew there was an unspoken rule: if you ran away from the control of one agent, the others would refuse to take you. It seemed they all knew each other and were interlinked in some ways.

As ever, Jawad looked to me to make a plan. "So, Gulwali, what do you say we do?"

I didn't have a clue.

FOR THE NEXT FEW DAYS WE FELT INCREASINGLY HELPLESS; then an idea came to me. I remembered the nice shopkeeper in Istanbul—from the DVD shop. He might help us.

We went to see him. He was kind and helpful, and gave us contacts for some other Afghans, people he thought might know of or were connected to various smugglers around town. Wherever they are in the world, Afghans still operate a tightly knit community. It took several calls and passing on of contacts to finally reach someone who sounded genuinely connected. But he explained he couldn't help us without a prior agreement from Amiri. He said no agent would do so, they operated a gentleman's agreement to not steal each other's people.

In the end, we did convince one guy—named Shir Aga—to help. A Pashtu from Nangarhar, Shir Aga was calm and considered sympathetic to our plight. He offered to speak to Qubat on our behalf and persuade Qubat to give him the money, instead of Amiri. But he warned us Amiri may still take revenge.

"If I take you, this is a big favor. This isn't how we do things here in Istanbul. Do you understand how risky it is for me to take you?"

I think Shir Aga knew Amiri was unreliable, because after a few moments of hesitation he let out a big sigh and said he'd take us. We were lucky. Shir Aga broke ranks on our behalf and told us to go to an address on the other side of the city.

As we made our way there, Amiri was constantly trying to ring us on the phone he'd given us. We tried switching off the mobile, but every time we turned it back on he rang again. In the end, we threw away the SIM card, but held on to the handset.

The address belonged to one of Shir Aga's friends, a nice Afghan man called Nour. Like Amiri, Nour was a Hazara, a musician and artist who'd been living in Istanbul for years. Unlike Amiri, however, this was the first time he'd ever sheltered refugees. He was clear it was a one-off and a favor to Shir Aga, but he seemed relaxed and laid back about it. He treated us like old friends.

Nour's apartment was small—just a single bedroom, a sitting room, a bathroom, and a kitchen. But there was a television, and it was comfortable, easily the best place so far on our long journey.

We stayed there for a week. Then Shir Aga arrived. He started by reminding us what a risk he was taking by allowing us to leave Amiri. But we were relieved to learn that Qubat had agreed to the deal and had paid him for our stay in Istanbul—as well as the next leg of the journey to Greece.

"I'm working on a plan," he promised. "A guaranteed plan. *Insha'Allah*."

Why was it that whenever an agent said something was guaranteed, I knew something was about to go wrong?

CHAPTER 21

We were sitting on the floor watching a film one evening, when Shir Aga turned up, beaming a big smile. "Good news. The game is direct to Greece."

"How?"

"I'll tell you on the way."

As the three of us got up to get our stuff, he pointed at Jawad. "You stay back. Only these two for now."

Jawad was furious: "I am coming too. You will *not* leave me here."

"Look, the boat is already overcrowded, and they only agreed to take the small two. But if they make it safely, you are definitely in the next game."

Jawad was still not happy. "No. I want to come." He looked at both of us imploringly. "Don't leave me alone. I want to stay with you."

I gave Jawad a guilty look. I liked him and I didn't want to leave him there on his own either, but at times like this survival trumped friendship. The game dictated that if one player got a chance to move forward, then he took it.

I was reminded of poor Tamim left with the parents of the smuggler with the sports car in Iran. Was he still stuck there?

"You'll go next time. Shir Aga has guaranteed it." As I said the words I knew how weaselly they sounded.

Mehran and I went with Shir Aga to his car, a large BMW parked in the street. I had yet to meet a regional agent who didn't have a nice car.

We drove back into the city and onto another industrial estate, where we were led into a factory building. Inside were rows and rows of tables with sewing machines on them. Sitting at the tables were refugees, mostly Afghan men and women, making Western-style dresses.

I suddenly wasn't convinced Shir Aga was telling us the truth about going to Greece. "Shir Aga, are you being honest with us? Are we definitely going in the boat?"

He wasn't annoyed by my doubts. "Trust me. Even my own brother will be going in the boat to Greece with you. You don't think I would betray my own brother, do you?"

This news did reassure me. But, if it were true, I had yet to see any evidence of a brother.

Before I knew it, a different man was ushering us into the back of a white transit van, which was already loaded with ten other men.

"Gulwali!"

As I searched for the voice my heart soared to see a familiar face—our old friend Baryalai. I couldn't believe my eyes.

"Gulwali." He looked at us, hurt, shaking his head.

"You guys left me in prison. What kind of friends are you? Because of you two I got myself into big trouble. I kept asking what was happening to you two, and because of that they suspected I was your agent. I got sentenced to six months in that bloody prison."

"Oh, no, I am sorry." I felt awful for him.

He continued: "I know you told the police officer I was your agent." He held up a hand in the face of my denial. "Don't lie, the police told me. You two deceived me."

Mehran and I were shocked he could think this. "No, no. We didn't. We promise."

"Don't lie."

"We didn't. We promise we didn't."

At this, Baryalai couldn't contain himself any longer: he burst out laughing, giving us both a big bear hug. "I know you didn't. I was teasing. I'm so proud of you both. Tamim told me all about the prison escape."

"Tamim? How do you—?"

A second voice spoke up from the gloom in the back of the van. "You little traitors left me as well."

Tamim was there too. I couldn't believe it. Two of my friends, right there in the van.

Before I could greet Tamim, Mehran starting interrogating him: "How did you get to Turkey? Why are you with Baryalai?"

Tamim brushed away the questions. "I walked. But enough of your questions. Where's Jawad? You left him too, right?"

I looked at him guiltily. "Er, yeah."

Tamim rolled his eyes and shook his head in mock disgust. "No loyalty." But his eyes were smiling, we all knew how the rules worked.

As usual, I had no idea where the van was taking us, but at that precise moment I didn't really care. I couldn't have been happier to see Baryalai and Tamim again. Friends were a rare commodity on this journey.

NEXT STOP IN OUR TOUR OF ISTANBUL WAS AN OPEN-SIDED shelter tucked away in woodland. It was vile. Not far from it was an Afghan-style toilet. Judging from the smell of this spattered latrine drop, thousands of people had come and gone without it ever being filled in or cleared. The stench was overpowering.

Including the twelve from the van, we numbered around 130—mainly Afghans, but with a smattering of Pakistanis and Iraqis. Baryalai introduced us to some of the men from the van: Engineer was a clever, geeky type of guy in his early twenties; Ahmad, who the others teased because he had some kind of skin disorder and spots all over his hands, was nicknamed *Gerasim,* "germs." There was also Hamid, a hugely charismatic, good-looking young man.

No one was happy to be in that shelter: it felt way too exposed and, as was so often the case, there was no food or water. Nothing made sense. Shir Aga had promised we were going on the boat, so why were we still on the outskirts of Istanbul in yet another filthy shelter? And where was Shir Aga's brother?

There were a couple of Pakistanis guarding us. They told us their boss would come soon.

This new agent introduced himself as Yassir. Instinctively, none of the Afghans liked being controlled by a Pakistani—our countries have long had an enmity and difficult history.

Mehran couldn't hide his distaste. "Why would an Afghan hand us over to a dal eater?"

It felt like a humiliation.

Yassir tried his best to convince us, and was polite. "I understand your concerns. This place is not appropriate for a long stay. We were supposed to leave tonight but there has been a problem and we must stay here one day. You Afghans are always in a hurry—such impatient people. But please let me assure you your worries are unfounded. I am working in your interests."

Everyone began trying to talk over him at once. He raised a hand and asked us to nominate one person to talk on our behalf.

"Baryalai," I whispered to my friend, "you do it."

Baryalai was just about to stand up and say something, but before he could get to his feet, Hamid had begun speaking to the guy in fluent English: "We were promised to go straight to Izmir. We demand to travel tonight."

The agent and Hamid conversed in English for a while. I was really annoyed—Baryalai was our spokesperson; this Hamid guy was half his age, not much older than me. Who did he think he was, speaking like this? And showing off with his English skills, too.

After he'd finished, Baryalai added a couple more points.

Yassir's response was curt: "You will leave when I say it is suitable. Why are you Afghans so angry? I gave you my word, didn't I?"

A few people continued to complain.

"Fine," Yassir finally said. "If any of you want to leave, be my guest. Go. There are too many people here anyway, which makes it dangerous for me." He looked around at us in exasperation. "Or you could be sensible and wait until the morning when, I guarantee you, you will be on your way to Izmir."

Another day, another guarantee from an agent.

But this time, at least, he hadn't lied. After a sleepless, hungry night, the Afghans were loaded onto four different vans, thirty people in each. I was crushed into a tiny corner in the back, my body pressed up uncomfortably against the metal. The ten or so Iraqis and Pakistanis stayed behind. I assume they were given different transport.

We traveled most of the day in the darkest, hottest, most confined space imaginable. It was hell, and I honestly thought I might suffocate to death it was so hard to breathe. We were thrown out into a valley at one point, where we rested for a few hours.

In the evening four new men arrived. One, a black-haired man with crazy-looking eyes and a bushy mustache, said he was our new agent. The other three worked for him. All of them were Kurdish. Together, they walked us through the trees to the top of a cliff.

From there, I could see the sea.

WE WERE EXHAUSTED AND THIRSTY, IN NO FIT STATE TO TREK for hours. Worse, the sun had just set, so we'd be slipping and sliding our way through a pitch-black forest. But what choice did we have?

As we walked, I tried to grab on to branches for support, but thorns ripped at my hands and made them bleed.

It was around 3 A.M. when we finally reached the edge of the forest and started making our way down a small, winding track, which led to the Aegean Sea. Out in the open air, I was reassured by the bright half-crescent moon in a cloudless sky. As we snatched a few minutes' rest I stared up and, as I often did, wondered if my family was seeing the same moon that night. Were they also thinking of me? And what of Hazrat? Had he passed this way too? Had he looked at the same moon from this same spot?

Tears sprang to my eyes. The past few weeks, since coming back into Turkey, had all been too much. I'd been on the move so often I really wasn't sure I could keep going. "Stop it, Gulwali. Don't be soft," I chastised myself. But I couldn't quell my overwhelming sadness: every particle of my being wished I was near home and hearth, food and comfort.

Wearily, I got to my feet with everyone else.

As we got closer to the seashore, I could occasionally make out the shape of moored boats, bobbing on dark water, all inky blackness, the size of it terrifying.

Hamid, who was talking with the black-haired agent, relayed that we were supposed to walk down to the sea, take off our shirts, and get in the water when the time came.

We made our way down onto a small beach. As we waited there for three or four hours, we saw a police patrol boat. I sat in my dark hiding place, watching the red and green navigation lights go past. A piercing spotlight on the bow probed the shoreline for signs of life.

I listened to the terrible crashing of the waves on the shore.

Even in the dark, Hamid could tell that I was scared of getting in the water.

"Relax," he assured me, "I will take you on my shoulders. I am a good swimmer."

Just as I was starting to feel better, the lights of a police car flickered in the distance. Someone shouted, "Get down."

The agent looked panicked. We all crouched behind a wall. The car slowed, but kept on going. One man near me whispered that we should walk in the same direction as the car, rather than risk this terrible water.

"Shut up and do as I say," hissed the agent. His manic eyes had a way of piercing through the darkness.

No boat came that night. Before dawn, we walked back up to our vantage point among the trees, cold and hungry, and exhausted from lack of sleep and spent adrenaline.

CHAPTER 22

ONE THING ABOUT LIFE ON THE RUN IS, JUST WHEN YOU think you're as miserable as you'll ever be, life manages to show you more, laughing even yet more loudly in your face.

I shifted my head against the rock I was using for a pillow—it was a waste of time trying to get comfortable, but it was better than the ground, where insects might climb into my ears. A tree root jabbed me in the small of my back. I swallowed my frustration and tried to roll over, but accidentally knocked into one of the sleeping bodies lying near me.

My thin jacket and jeans were no match for the clammy, damp earth beneath me, despite it being a sunny afternoon. I reached into my little bag and pulled out a T-shirt, covering my face to block out the light.

"Gulwali, stop moving about. You'll wake them all up."

"Sorry," I whispered back to Baryalai. "I'm cold."

"I know. Try to rest. Don't make these guys angry, okay?"

He was right to warn me. Tempers had been really fraying. We were all desperate, hungry, and increasingly angry. The nights were the worst. It was always hard to sleep. But, fearful as I was of the trees and unknown landscape, I was more fearful of the other men.

Since returning from the seashore, the Kurds had just disappeared and left us there, for three uncomfortable days and nights. People were at their most cold and tired, so almost anything could set them off. Even if someone said something nice, the others shot him down. We were huddled in a small clearing, and there were a lot of fights about who would get the softest bit of ground.

During the daylight hours, we slept on the ground with only the trees for shelter; at night it was too cold to even attempt sleep, so instead we played cards. Tamim, Mehran, Baryalai, and I huddled together for warmth. The only comfort I had now was being reunited with these friends, especially Baryalai, who had always looked out for me.

Since talking to Hamid that night at the sea, I had grown to like and trust him very much. He, Ahmad, and Engineer had a bond, much like my little group, but we all became kind of allies.

Everyone was too miserable to talk much but we gleaned little bits of information from each other. It turned out that Tamim, Ahmad, and Engineer had been briefly kidnapped by our old nemesis, the drunken bully Amiri. Our friendship was almost undone when we worked out that they'd paid the price for our running away: it seems Qubat had refused to pay him and had refused to send any more refugees his way. By way of revenge, he had held the three of them hostage for a few weeks.

Tamim didn't see the funny side. "Not only did you little shits leave me on the border, you got me kidnapped too. If the Taliban don't kill you, I swear to God, I will."

Mehran and I thought it was hilarious.

"We were just lucky," Mehran teased Tamim. "Stick close to us next time, and we'll keep you safe."

I couldn't help joining in: "You bring us bad luck. Because of you three, we are stranded here. As much as we like you, are you guys really to our advantage?"

IT WAS JUNE 2007. I HAD LEFT MY HOME IN OCTOBER THE year before, making me an undocumented refugee, an illegal, for nearly eight months now. Only now was I getting closer to Western Europe.

At no point in this journey had I any idea how to go about claiming asylum, even if I had wanted to. None of the three countries I'd been stuck in for the past half-year—Turkey, Iran, or Bulgaria—had felt safe enough for me to want to stay. And I clearly was not welcome in those countries: prison or deportation was all they could offer me. I knew I had to keep going, keep moving—and hope that the promised lands ahead did actually offer safety and security.

I was glad my mother would never have any idea of the times I had felt so lonely or cried so hard and so often my eye sockets ached—full days lost to tears and headaches, sorrow and emptiness. But I think, on that cliff top, even among my friends, I started to feel sadder and lonelier than ever before. As I struggled to sleep on the hard ground, with a growling, empty stomach, I couldn't help but let the memories flood in—memories I usually tried to block out. I'd dream I was back at home, teasing my cousins or listening to my mother hum softly as she folded our laundry, the house filled with morning light and smelling of freshly baked naan—only to wake cold and hungry to yet another lightless sky above the forest canopy.

What scared me most right then was that we'd been there for three days. The Kurdish agent with the manic eyes and his trio of accomplices had failed to return. The agent had thick black hair covering his hands, which matched the hair on his head and the little tufts poking out of his ears. Between the hair, the mustache, and those eyes, he reminded me of a spider.

Mehran and I decided to find out if Shir Aga's brother really was among us. We began by trying to find someone who looked like him—no one did. But we were so hungry, our vision was blurred anyway.

People tended to congregate in small groups. As a child, it was easier for me to go and talk to others: "Excuse me, I am looking for the brother of Shir Aga."

No one seemed to know him.

Later, I was talking to Mehran when a young, green-eyed man with red hair approached us. "Why are you looking for Shir Aga's brother? What's your business?" His tone was tough.

Baryalai noticed and bristled. "Little man, is everything okay with you?"

I think I knew this was our guy. "It's no problem. I know this man."

Baryalai still looked suspicious. "Then sit, my friend, and join us. Any friend of Gulwali's is a friend of mine."

The young man sat down and spoke in an urgent whisper: "Look, stop asking questions about me. If some of these guys know who I am, it will make big trouble for me."

That was true. Tempers were fraying so badly that if the others knew the connection they'd surely beat him up.

I used this to threaten him: "If you don't do what I say, I'll make a public announcement. You need to contact your brother and find out what is happening."

I noticed Baryalai giving me a surprised, but slightly admiring glance. I'd grown up a lot since we'd last been together.

The young man looked desperate. "I don't have a phone."

"So find one."

We'd been warned by Pakistani Yassir not to carry any mobiles with us because the police could use them to track us down. The only time I'd ever had a mobile in my possession was briefly in Istanbul—the handset we'd acquired from Amiri. But we had left that with Jawad back at Nour's apartment.

Somehow, the young man managed to find one and call Shir Aga. He came back with a slightly arrogant air: "My brother says to relax. He's working on the plan. It's guaranteed. There is a new boat, and he's negotiating with the captain as we speak."

I hoped he was telling the truth because, by the next day, people were beginning to think we'd been abandoned. There was talk of leaving, and everyone was arguing with each other. Some people wanted to try to walk to a road and find help; others told them not to risk being seen and arrested.

I knew for certain that turning ourselves in would be a huge mistake: "Do you want to go to prison? Some of us here have been. It's no joke."

An older man raised a fist at me in warning. "You know nothing, boy. Shut up. Children don't speak over their elders."

I knew the people who didn't know me personally wouldn't listen to me because they only saw me as a child. So I figured it was better to let my older friends argue it out. I knew Hamid and Baryalai would make good decisions on our little group's behalf: I prided myself in making sure I made friends with the smart people. But I think, deep down, we all knew that if the spidery-looking black-haired Kurd didn't return by day four or five, we'd have no choice—it would be starve or surrender to the police. Every hour that passed, I prayed and hoped he would return as I tried to quiet the rumbling pain in my stomach and quell my nauseating fear. On the third day, about eight in the morning, we were sitting in silence, still hoping for a miracle, when the agent and his men came back. They were carrying crates of tomatoes and several loaves of bread, as well as two large 10-liter bottles of water.

I was so hungry I could have easily eaten several loaves all by myself. The crates of tomatoes might have looked like a lot but after they were distributed, there was only one per person. While it was the best tomato I had ever tasted in my life, it did nothing to quiet the howling in my hungry stomach—just a pebble thrown into an empty well.

The black-haired agent watched us eat with a look of distaste. Filthy and starving, we snarled and whimpered like a pack of wild animals. For him, this was nothing new; every week he probably witnessed scenes like this, bringing in a new batch of dirty, hungry, desperate people—scum, in his mind. But we were the scum who would make him rich.

Finally, when the last crumbs had been sucked from beneath our filthy fingernails, he gave us the news we'd been waiting for.

We were going to Greece.

WHEN WE REACHED THE WATER'S EDGE, IT WAS CLOSE TO MIDnight. The Kurdish guides started to push people in the direction of a small speedboat.

Hamid translated for the smuggler as he pointed to a boat bobbing on the horizon. "Get on the small boat and it will take us to the big boat."

I didn't know much about boats, but even I could see that what Hamid called the "big" boat wasn't very big, certainly not big enough to fit 120 people.

None of us wanted to get into that water, or into the speedboat, or onto the so-called big boat beyond it. In the dark, it was impossible to make out how deep the water was or what dangers lurked beneath. Soon, though, the speedboat began ferrying people—as many as they could squeeze in at a time—out to the other boat.

My new group of seven friends hung close together but somehow, in the crush, everyone but Mehran and I managed to board. I began to panic, thinking we'd be left behind. But the speedboat came back. It was my turn.

"Get in," Mehran yelled over the din of the motor. "We have to get in. Come on, Gulwali."

My feet were stuck to the sand in fear. I could not get into that water.

Mehran literally dragged me into the sea and practically threw me into the boat. As the speedboat lurched to life, I nearly vomited with the strangeness of the sensation. I was absolutely terrified, but I also knew that if I could survive this, I was one step closer to Western Europe.

On the other side of this horrible water lay Greece.

"Stay in the game, Gulwali, stay in the game." I tried to control my emotions by talking to myself.

When the speedboat reached the main boat, I was instantly glad we had been one of the last on board. The first arrivals had been shoved into the claustrophobic hull below deck, while the others were sitting on a wooden bench that ringed the main deck, which was also crammed with people. I could see Hamid sitting close to the steering wheel. Standing over him was a mustachioed man I assumed to be the captain.

Mehran and I didn't know where to go: there was no room anywhere. Suddenly, a young crew member, a pale-skinned teenager with pierced ears and a tight T-shirt, grabbed us both. "You," he ordered, pointing Mehran to the galley steps. "Downstairs." I went to follow. "No. You, the small one, stay on top."

I managed to squeeze past the bodies to get to where Hamid was sitting.

As the final people continued to climb aboard, we were sandwiched in tighter and tighter together until I could no longer move my arms. Breathing in the oniony smell of the men, I began to feel nauseated, and it took all my self-control not to vomit over my boots.

The captain, a burly-looking, muscular man, stood by the steering wheel close to me and Hamid. He had an intense look on his face and shouted at the Kurd in Turkish. Hamid translated for me in a whisper, pulling comical faces at the swear words: "You told me we had sixty tonight. Why have you sent me so many? We can't take them all. This is a fucking pleasure boat, not a ferry."

"Fuck you. Do you want the money? This is the cargo. Take it or leave it."

The Kurd argued and argued with the captain, who demanded that half the people get off. Many on my deck looked like they wanted to. The captain had locked the hatch to the hull and those inside, including Baryalai, Mehran, Tamim, Ahmad, and Engineer, were starting to shout, demanding to be let out.

But the Kurd was having none of it. He took a wad of cash and handed it to the unsmiling captain, who huddled with the teenaged crew member.

After the Kurd sped away again across the water, the captain started his boat's engine, which coughed to life, sputtering out a greasy cloud of smoke. I was sickened by the crowd of bodies, the smell of diesel, and the rocking of the boat, which seemed as though it might capsize at any moment.

I said a silent prayer, begging Allah to spare my life.

Just as the boat began to make progress, and just as we were beginning to settle into our voyage, chaos broke out. The boat suddenly became enmeshed in a floating barrier. It was like a fishing net made of barbed wire. The weight of the boat must have triggered some kind of trip hidden deep in the water.

Whatever it was, it alerted the coastguard: in the distance, I saw flashing lights and heard sirens.

The captain wrenched at the steering wheel, pushing the boat back and forth, while the crewman yelled at us to keep our heads down.

Somehow, in the dark, the captain found a way out of the mass of wire and, through the clattering roar of the engine, the sirens slowly began to fade.

WE WERE IN THE MIDDLE OF THE SEA.

The captain had turned off all the lights on the boat to avoid detection. I could make out nothing but endless black. I heard only the dreadful hiss and crack of waves slapping and sucking at our overloaded vessel. There was no joy at our escape, or sense of new hope. It was terrifying—like landing on an unknown planet.

The people in the hull started to bang on the door again. They were trying to move around, shouting that water was coming in below. The captain took out a gun and fired it into the air, which quieted their protests. Again, I said a silent prayer, this time in thanks that I wasn't sitting down there. But I said another prayer for my friends who were trapped below.

After a couple of hours, land began to appear through the hazy dawn. It had to be Greece. I looked with blurred eyes, not having slept properly for several days, just staring out at the glimmer of land, listening to the hushed conversations of those around me.

The captain nosed the boat toward a small, rocky beach.

Someone called to the people below, "Greece. We made it to Greece." They shouted back their praise to Allah.

I shoved my way through the crowd to the side of the boat. People were already jumping from the deck, picking their way over the rocks to the shore. The teenaged crew member had unlocked the hull and was trying to quell the panic as people clambered up the stairs and off the boat.

Baryalai reached the top deck and grabbed my arm. "Get ready to jump, little man. Stay with me."

On my other arm I felt Hamid's hand. "Something isn't right," he said. "Wait."

It was impossible to work out what was going on, or what to do. People were pushing and jumping, and the boat was beginning to lurch.

As people continued to pour off the boat, the teenager started pushing everyone back down the stairs and into the hull.

"Let us get off."

"What are you doing? We're in Greece."

The captain started up the engine again, chugging the boat back into the sea. Everyone started screaming: "Let us off! Let us off!"

In a panic, one man tried to climb off the side of the boat, but the captain clubbed him with his gun, then waved it menacingly toward the rest of us. "Sit down," he warned.

I was horrified, shocked even more to see that, in the scramble, Baryalai had somehow gotten off the boat—without me.

Hamid put his hand on mine. "We'll be okay, Gulwali."

"I want to stay with Baryalai," I said.

"Gulwali, stop it. He'll catch us up. He'll be fine. And so will we be. Really."

As we moved further out to sea, my heart began to sink. I glared angrily at Hamid.

"I told you it wasn't right, Gulwali. I don't think that was Greece."

"Why didn't you warn Baryalai? Why did you let him get off?"

"I couldn't stop him. I wasn't really sure. I'm sorry, Gulwali."

Tears burned my eyes. After once again finding Baryalai, I couldn't believe I had lost him again. He'd been my protector for so long.

Not only had I lost Baryalai, but I didn't know whether Mehran and Tamim were still on the boat or if they too had jumped onto the shore.

WE FLOATED FOR TWO DAYS, SEEMINGLY DRIFTING TO NOWHERE, although we'd been told the journey would only take a few hours. My whole body hurt from lack of sleep and the painful position I'd been sitting in.

My stomach ached. We hadn't eaten since the first day, when the captain had given us a few pieces of bread.

I was sure now I would die. Everyone expected the captain to throw us overboard. I would drown in this murky water: an icy, cold, lonely death—away from my mother's warmth, my father's strength, and my family's love.

The journey was supposed to be the beginning of my life, not the end of it.

"HELP US. OVER HERE. HELP US."

I began to hear foreign voices over our own. I dared to look up, and couldn't believe my eyes. Four coastguard boats were circling our stricken vessel. People were cheering, shouting, "Over here. Over here."

My first, fleeting reaction was one of disappointment. I didn't want to be arrested and sent back, especially not to Turkey—again. But that thought passed in seconds—I was only thankful I wasn't going to die.

A voice from a loudspeaker boomed at us, words I couldn't understand. The police on the nearest boat were gesturing wildly. Everyone was screaming at them. Some men leaped into the water, making the boat rock dangerously from side to side even more.

I hesitated. I didn't know if these people were lying to us too: if they were going to help us or if they'd let us die. But the boat was going down fast. If I jumped in I might drown, but if I didn't I definitely would drown.

Preparing to jump, I held my breath and closed my eyes. The waves were so big. I wondered how long I could stay afloat by flailing my legs. Then something fell beside me; it was a rope. The coastguards were throwing ropes up to the ship. Desperate hands lashed them to deck cleats as the police boats managed to lasso their boats to ours to keep us afloat.

The coastguard began fishing people out of the water, throwing life jackets to those of us still on board. I had no idea how a life jacket worked, but I yanked it over my head, securing it across my chest. And it was then that I realized, for the first time in days, that I felt safe. My body shook uncontrollably and my legs buckled beneath me as shock set in. As our boat was righted and towed to the shore, shock seemed to spread through our crowd like a virus. Grown men cried openly and shook with relief, thanking Allah, whooping and cheering, celebrating that we were alive. Hamid hugged me with great joy.

The police tied our boat to a large metal hoop fixed into the solid stone of the harbor wall. We were told to stay on board and not to move. I wanted so badly to get off, but I took solace in the fact that at least the boat was still afloat, but no longer going anywhere.

A few people boarded, one of them a smartly dressed woman with flowing light-brown hair. She came over to me and threw a blanket round my shoulders, then handed me a bottle of water. She had translucent olive skin and round blue eyes. She was wearing a fluffy white sweater and, in my delirious state, she looked just like an angel. She wasn't there with the police or the coastguard, but with a group of townspeople who had heard we had been brought ashore. This I couldn't believe. It was the first act of human kindness I had witnessed in weeks. As the woman fussed over me, saying words I didn't understand, I started to cry: all of the pain and fear and loneliness of the past three days came out in great, big, hiccupping sobs.

The woman's face crumpled sympathetically, and she motioned to a friend, who brought over some fish with vegetables and a bar of chocolate. She smiled at me with her big kind eyes and continued to speak to me in her soft voice. I stared at her with tears rolling down my face, shoving the food into my hungry mouth. I was so dehydrated, I could barely swallow. But, once I'd forced down the food, my shrunken stomach couldn't handle it, and I immediately needed to go to the toilet. I screamed at the coastguard to make them understand that I needed to go—*now*. Two men escorted me off the boat toward a brick toilet block. But they stood guard, watching, though thankfully, they were decent enough to wait a discreet distance away as my bowels exploded.

When I returned from the toilet, I was doubly relieved to see Mehran, Tamim, and Ahmad were all there, safe, though as shaken and upset as I was.

Over the next twenty-four hours, the coastguard recorded everyone's name, age, and country of origin. Hamid, again putting his language skills to good use, translated their requests for us and relayed our answers back to them in English. After everyone was processed, we were loaded onto a ferry and taken to the other side of the harbor, where two coaches waited to drive us to Athens.

CHAPTER 23

I FELT AS IF I WERE ENTERING A FORTRESS. GUARDS WITH guns and fierce-looking dogs stood behind a 3-meter-high steel-mesh fence topped with concertina wire; more guards stood by at several security barriers we had to pass. For the fourth time in six months, I was back behind bars.

We were fingerprinted, and the police brought a doctor who gave me some medicine for my stomach. A Pashtu-speaking Afghan working at the center translated the police's interrogation sessions. They asked us our names, our ages. They asked again and again about the captain of the boat, if we knew where he was. I just said the same thing, over and over: "I don't know."

A few days later, I found out that the authorities had arrested him. As the boat went down, someone had handed him a *shalwar kameez* to wear so he could blend in with us. But, of course, he couldn't hide his ethnicity during interrogation. A police officer told me he could be jailed for twenty-five years. It's very hard to say how that made me feel. He was paid a lot of money to take us to Greece and, by agreeing to overload his boat, he nearly killed us. But I forgave him because we survived and that at least was something to be grateful for. I also felt sorry for his family, who would miss him if he were in prison.

I told the translator about Baryalai and the others left behind. He promised us someone would investigate, and that if they had been dropped on a Greek island near Turkey, it was likely they would have been deported back to Turkey.

I felt terrible. Baryalai had been there for me so many times, always sticking up for me, always protecting me, coming back for me, urging me to carry on, calming me with his words of wisdom. If I'd had one true friend on that whole awful journey, it had been him.

But, as tough as it sounds, I resolved to move on and forget him. I had no other choice. Those were the rules of the game: keep going, don't look back.

I'D HEARD WHISPERS FROM OTHER REFUGEES THAT WE WERE going to be deported back to Turkey.

I was terrified that I could be sent back at any moment. It would have been a bitter blow. I'd finally made it to Western Europe and I did not want to be sent back. I was still living the game of Chutes and Ladders. This was how my life was for now—I accepted the pitfalls and perils, but that didn't make them any easier to bear.

As ever, getting information on our legal situation was a constant source of stress and confusion. Officials at the immigration center gave us papers detailing our case, but what were we supposed to do with them when they were written in Greek?

We were questioned again and again.

Because Hamid spoke the best English out of all of us, the police conscripted him as our group's translator.

At one stage, the police interrogated a frail old man. He was weak and trembling, and I don't know how much he actually understood about what was happening to him. His answers certainly got Hamid into trouble. The policemen were asking questions of the old man. Hamid translated their questions to the old man, then translated the man's answers back to the officers, who didn't like what Hamid was telling them.

"How old are you?" Hamid translated to the old man.

"Twenty-three," he answered

"What is your final destination?"

"Europe."

"What is your preferred country in which to seek asylum?"

"Europe."

The man was obviously getting his numbers confused and didn't know where in the world he was, but the smaller of the two officers thought Hamid was being insolent with the answers. On the third answer, he whipped his hand across Hamid's face.

We sat up in shock. The other officer yelled something angrily at his colleague.

That night, the guards put us in one large holding cell. It was incredibly cramped and we were all filthy; the smell was horrible. They fed us, though our meal was lukewarm, served on a plastic corrugated tray. When we lay down to sleep, we sprawled over each other, each man's body forming a pillow for another. I slept with my head on Hamid's legs.

The following morning the guards came and separated eleven of us from the others: Hamid, Ahmad, Engineer, and I were part of the smaller group. I assumed we were going to be questioned again, so I didn't bother saying farewell to Mehran and Tamim. But, instead of leading us along the now-familiar corridor to the interrogation room, we were taken to a different wing, where they issued us new clothes. We also got towels and toiletries, and were finally allowed to take a hot shower.

I asked one of the officials about Mehran, and I managed to pick up some information through my basic understanding of English. It seemed we had been separated according to age. But this didn't make sense to me, because Mehran was younger than Hamid. Mehran was tall for his age, so perhaps that's what had swayed their decision.

Baryalai was gone, and now I was without Mehran too. We'd been together since those very first days in Mashhad, Iran. He made me laugh, and could always be counted on to lift my spirits when I needed cheering up. It was hard to rely on these friends for so much, and even harder to have them taken away from me.

And of course my brother was still out there somewhere too. Thinking about him, where he might be or what difficulties he was going through, was like a raw, gaping wound in my mind. But at least I would still have the support of the three new friends I had made in Hamid, Ahmad, and Engineer.

When we arrived on the wing we were put in separate cells, but after a day we proved we could behave well, so the guards relaxed a little bit and let us share rooms and have our meals together. Hamid and I became cellmates, while Ahmad and Engineer were just a few doors away.

When you are traveling, life becomes all about survival. Days and dates are of little importance when food, shelter, and personal safety are the daily priority. But, one morning, we realized it was September 12—the first day of Ramadan, the Islamic holy month. The warden agreed to let us fast during daylight hours, as is the requirement during Ramadan, and we were allowed to eat our meals at night, when food is permitted.

I was greatly touched by this. Greece isn't a Muslim country, and even though we were in a detention center, the staff went out of their way to make us comfortable. It was very kind.

A group of young Albanians was also being detained in our section. They were loud and noisy, and weren't allowed to eat their meals together. The guards seemed to take great care to keep them locked up. It didn't go down well with them that not only were we allowed to eat together, but we got special privileges too. I think we probably also woke them up when we were eating in the middle of the night. Each time we walked past their cell, they would scream and throw their shoes at us, or lunge at us if we walked too close. They were just bored, I think—picking fights and intimidating the Afghans seemed more fun than anything else they could do. A prison guard called Dimitri took it upon himself to stand up for us: he would tell them to shut up, and threaten to send them to the adult wing. I think he had real sympathy for our plight—as though he understood our pain and was genuinely concerned.

He became a regular and welcome feature of our day. Dimitri's English was good, so thanks to Hamid we could share stories of the lives we'd left behind in Afghanistan. He was genuinely interested. We were talking one day, and I told him about the events leading up to my journey. I couldn't help becoming very emotional and crying, while he listened carefully, his face etched with sadness. It felt good to talk to someone about what I had been through.

We showed him the case documents we had been given so he could explain them to us.

"Are we going to be deported to Turkey?" asked Hamid. This was the burning question.

Dimitri stared for a time, his eyes dancing over the paper, then he threw his hands in the air and made a noise like he was exasperated.

Hamid translated: "Bureaucracy. It's idiotic," he said.

"But what does it say?" I replied, desperate to get to the bottom of the paperwork.

Dimitri was speaking again.

"He says this is a permit allowing us to stay in Greece for one month. After that, you must leave the country or face deportation," translated Hamid.

We shrugged. It could have been worse.

Dimitri left the cell promising to check our files to see if there was anything further he could do. The next time he was on duty, his face told me that we were in for bad news.

"He says we are sentenced to three months in jail," said Hamid.

"But we haven't been to court," I protested.

"It seems we were sentenced anyway. He says the court found us guilty of entering Greece illegally, and we need to stay in here for our own protection because we are young."

"Protection? From what? The crazy Albanians that want to beat us up the first chance they get?" I was incensed. "And after three months, what then? Can we stay in the country?"

Hamid and Dimitri resumed their conversation. It was frustrating doing it this way. I concentrated hard to see if I could pick up any of the English.

"He says he doesn't know. That we need to talk to a lawyer about that."

I was really angry now. "You got a lawyer? Anyone? A lawyer going spare? Check under my bunk, I think I might have left one there."

Hamid was more thoughtful. "This doesn't make sense. Three months in prison, but the other letter says we have to leave Greece in one month."

Dimitri said something.

"What did he say?" I demanded. I was cooling down a bit now—my mind beginning to focus on the problem at hand.

"He says it doesn't make sense."

"I know that. Why are you telling me something I already know?"

We were just going round and round in circles.

MY HEALTH WAS VERY POOR. MY STOMACH HAD BECOME DIStended through malnutrition, and my body was covered in pimples. At night, my sleep was filled with terrible dreams and flashbacks. I used to dream I was drowning, or wandering lost in the mountains. Sometimes I would wake just as I relived that terrifying leap from the moving train in Bulgaria. Night after night, I would wake up shaking, screaming, unaware of where I was. My nightmares were common knowledge:

when they happened I woke the whole cell block up with my screaming, so I was taken to see a psychiatrist.

She was a soft-voiced woman in her late thirties. Once I would have been horrified to sit in the same room and talk to a woman whose head was uncovered. But I had changed. Some things really weren't that important to me anymore. And the nightmares were so bad that I knew I needed her help.

Hamid came with me to translate. "She says to tell her about your dreams."

"Well," I said, talking to Hamid. I found that strange, as though he was the psychiatrist, so I turned toward her instead. "They happen almost every night. The dreams end the same—I wake up and I think I am at home, but I am screaming for my mother."

She nodded the whole time I talked, even though I was speaking in my native Pashtun. She listened for about an hour. I'm sure it was as frustrating for her as it was for me. Hamid did his best, but it's very hard to talk freely and openly in a situation like that. Not that I had huge expectations of her: I was a twelve-year-old Afghan boy—in my culture, we don't go to psychiatrists. If something like that happened you'd either keep it quiet and tell no one, or go and see a religious scholar instead.

In the end she prescribed me some sleeping pills. They helped me sleep better, but I could still feel the demons in my head. A few pills weren't going to magically make them silent.

Dimitri arranged for us to see another doctor about our physical health. Poor Ahmad's skin disorder was really bad—his hands looked as though they were falling off. The doctor gave me some cream for my pimples and some other medicine for my stomach. We wore a prison uniform, which was kind of like a tracksuit, but the doctor also gave us some secondhand clothes to wear for when we got out. That was very touching.

The best medicine, though, was talking to Dimitri. Even though Hamid had to translate, Dimitri always took the time to listen. I found

I could express myself freely to him, and he didn't mind when I cried occasionally. Sometimes he would tease me and rub his eyes, pretending to be me crying. The teasing didn't upset me; if anything, it made me feel more comfortable with him.

"I wish I could take you boys home with me," he liked to say.

That made me cry even more.

TEN DAYS INTO OUR DETENTION, DIMITRI CAME IN LOOKING very pleased. He immediately started talking to Hamid.

"Good news," Hamid translated. "He says the doctors have convinced someone senior to let us go to a UN refugee camp. Dimitri says there's a football pitch and it's much better than here."

He was only sort of right. A UN representative came to collect us from the prison. As we stood with our bags and new clothes in the car park, the UN official looked us all up and down, as if carefully checking us over. Then he turned to Dimitri and a second prison guard who was with us, and gestured to one of our group, a youth called Aman.

"I'm not taking this one. He's too old."

Aman looked shocked; we all were. We all started arguing and trying to say we wouldn't go without him.

Dimitri reasoned with us. "It was hard to persuade these people to take you. You others have to go. He has to stay here. I am sorry."

Reluctantly, the remaining ten of us went. I hugged Dimitri, who promised us he'd come and try to visit us.

The camp was in a rural area about forty-five minutes outside Athens. There were some large communal buildings with a kitchen and offices, then row upon row of small container cabins for sleeping in, as well as shower and toilet blocks. The four of us were able to share a cabin together.

It was comfortable, although I was aware that we had really only traded steel bars for wire mesh. The official had told us we were not allowed to leave the camp without his permission, and he made it clear that Dimitri had guaranteed we would behave ourselves.

Because we were still fasting for Ramadan, they gave us a choice of either having pre-cooked food or to be given a daily allowance of ingredients, which we could cook ourselves in a communal kitchen. We took the second option and shared the responsibilities between us.

Some of the residents—mainly Arabs and Afghans—had been in the camp for months and even years while they were waiting for their paperwork to be sorted out. Some were trying to claim asylum in Greece, while others were trying to claim asylum in European countries where they had family connections. They warned us that if we started the process there, it could take a very long time.

Although I knew I could get legal advice from the UN or claim asylum in Greece, I didn't want to get trapped anywhere. I had to find Hazrat. The others didn't want to stay there either. And everyone was worried about the documents saying we had to leave Greece within one month. We decided we couldn't risk being sent back to Turkey. Besides, no one had asked us if we wanted to claim asylum and we didn't know if we could or not. Without that security, there was no way any of us were going to put trust in the system. We'd seen enough to know there was little logic to what happened to refugees, and even less effort to communicate it.

We'd been in the camp for two days when Dimitri brought his wife to visit us.

It was good to see him, if only for a little while. He sat with us in our cabin and seemed really concerned about how we were. We confided in him that we wanted to leave. Dimitri tried to dissuade us against leaving, assuring us the UN would sort out any problems. I didn't believe this but, by the end of his visit, we promised him we wouldn't leave.

As we waved him off, he said he'd come and visit us again soon. But from the sad look on his face, I suspect he knew our pledges to not try to leave were lies said to make him happy.

WE HAD OUR ESCAPE PLANNED. WE WOULD GO TO THE NEAREST bus stop, from where it would be back to Athens and then on to Patras, a shipping port where trucks and semis embark for Italy.

Late that afternoon, we walked to the volleyball pitch at the far end of the camp, near the outside gate. After messing around on the pitch for a while, we jumped over the wire barriers—they weren't very big—and ran into the surrounding trees. No one tried to stop us.

We made for Athens. A member of our group had a relative there and he had managed to call him from the camp. The relative had told us where to go in Athens—a certain square where newly arrived refugees would congregate and get picked up by their agents. Once we got there, we were supposed to contact him.

I think I had about $150 left, but I had a feeling that was not going to get me far; indeed, when I exchanged my remaining money for euros a few days later, I was dismayed to see that in the new currency it seemed a lot less.

In Athens, we managed to find the square we'd been told about; there were so many Afghans congregating there it felt almost like a street in Jalalabad. Our friend contacted his relative, who told us to wait there for a few hours. It must have been obvious that we were newcomers, as all sorts of people approached us offering advice and help, such as arranging transport to Patras. We didn't trust anyone. The whole scene seemed so bizarre. Even more worrying was that some of these people appeared to have information on us, asking us if we were the people from the boat that had been intercepted, or the boys who had left the prison. One man even told us he knew we had just come from the UN camp. I could not understand how these strangers seemed to know so much about us. It was really unsettling. But I don't know why I was surprised, given how tightly I knew these networks worked.

One man came up to me and took me aside. Hamid stood in discreet attendance, keeping me under his watchful eye.

"Are you one of Qubat's people?"

I couldn't believe it. I was scared. "No. I don't know what you are talking about."

"So you don't know the name 'Shir Aga' then?"

At this I got even more worried. How the hell did this guy know all this? I clammed up completely.

He carried on talking: "I am here to help. I am a representative of Qubat. Shir Aga is my relative. His brother, who I believe you know, is staying with me."

This was beginning to make a bit more sense, but I was still wary.

"What happened to the others you were with on the boat?"

"I don't know. They were in the prison."

"Some of these men haven't paid their bill to Qubat. Their families want confirmation they arrived safely before they pay the next due installment. I need you to contact them and tell them to inform their families to pay."

Why was this up to me, I wondered. "I can't. They are in the prison. What can I do? I just got here. Please leave me alone."

His tone was threatening. He told me he needed to meet again here in two days, otherwise he would come and find me.

As I went back to Hamid, I realized my hands were shaking.

Fortunately, our friend's relative arrived and took us all to a flat nearby. We had food and tea there. His relative stayed with him, and we others had to find our own way. The man helped Hamid, Ahmad, Engineer, and me to find a cheap, Afghan-run guesthouse to stay in. It was populated by Afghan laborers. Their life was so tough. It was still Ramadan, and as soon as they had eaten *suhur,* while it was still dark, they would leave the building and go and wait by the side of a road to be picked up by locals in need of cheap daily labor, usually on building sites. They wouldn't return until very late in the night, exhausted and dirty. They earned only a few euros a day, barely enough to pay for their food and rent. Sometimes, even worse, they returned to this house a few hours after they left, disappointed, because no one had picked them up.

For the first time on this journey I wasn't an illegal: I still had two weeks left on the document allowing me to stay there one month, which allowed me to walk around Athens freely, unafraid. For the first time, I could relax and act like a tourist. Hamid suggested the four of us go get proper haircuts and hot towel shaves—I didn't have any hair on my face to shave but he, Ahmad, and Engineer did and I didn't want to be left out, so I insisted on having one too. We walked to all the famous sites and I distinctly remember going to Chinatown. I had never seen anything like it. For the first time in my life, I tried Chinese food. We found some cheap clothes shops and bought ourselves new clothes and shoes. My boots—my old best friends—had served me well for the last ten months, but they had finally fallen apart.

Two days later, even though I was really scared, I went back to the square to meet the man who had threatened me. He impressed on me that I do as he said: "Look, boy, don't worry about the others now. What is your plan?" He pointed at Hamid and the others. "Those people are your friends?"

"I don't know. But that's their business, not yours." I wanted to stand up to this man.

He told me that he'd been instructed by Qubat as well as Shir Aga, to meet me and explain to me the different options I had now.

"Your family has paid eight thousand dollars for us to arrange passage to Italy, including your expenses."

This was complete news to me. Until this moment, I had had no idea how much money had been paid, or for what. Now I realized why some agents had been giving me money along the way.

My family had paid all this up front. It was a huge amount. But all I could think about at that second was talking to them.

"Can you help me contact my family?"

"Do you have a number for them?"

"If I did, would I be asking you for help?"

"Don't give me cheek, boy."

I checked myself and tried to be a bit politer, but I still didn't like this man. "What about my brother? He was supposed to travel with me. I know my family wanted this. So where is he now?"

"I don't know about your brother. I was only informed about you."

He went on to say that because my family had paid up to Italy, he would now arrange for me to get to Patras, and then to Italy.

"You can leave tomorrow. It's guaranteed you will be out in a truck and in a ship. You will be in Italy in no time."

That word again: "guarantee." It made me very wary. And the idea of getting back on a ship made me too terrified to even think about it.

"Do I have another option?"

"Yes, boy. I can give you some money back instead, and you can make your own way, which you will, no doubt, regret."

I somehow doubted that. I definitely preferred to stay with my friends than to go with this creature.

"I'll go with my friends, thank you."

"Your choice." He sucked in his cheeks, fished in his pocket and handed me 300 euros.

I was shocked. "What's this? This isn't enough for me to get to Italy. There's a guy over there who charges one thousand dollars to go. You just said my family had already given you the money. Give me back what they paid."

He got very angry at that. "I came here with my generosity. Shir Aga said you were a kid and I had to take care of you, but you are rude. If you want to take it, take it—if not, your choice."

"But you said my family paid. I want their money back."

Hamid, who had been keeping watch again, could see things were getting tense. He walked over.

"Give me the money. My family paid Qubat. I want their money." I was becoming hysterical.

"Get out of my face, boy."

Hamid was at my side now. "Is everything okay here?"

"No. This stupid kid thinks he can take Qubat's money. We are generously giving him three hundred euros back, but he doesn't seem to want it."

"But it's my family's money," I persisted.

Hamid grabbed my arm and turned to the man. "I'll take care of him. Thank you, sir." With that, he took the notes from the man and dragged me away. Once at a safe distance, he told me off: "I thought you'd be a bit wiser by now. You don't argue with these people. He came to find you, to give you money, and you fight with him? Let's get out of here."

That evening we got a train to Patras, but when we arrived, we found the conditions terrible. There was a glut of refugees living on the streets, under bits of plastic sheeting or in tents made from branches. They cooked in the road. There was no clean water to drink and even less to wash with. It was raining the night we arrived and mud was everywhere, even where people were trying to sleep.

It was now the fifteenth day of Ramadan, halfway through the month-long fast. I had really wanted to continue fasting but there was no way it was possible in these circumstances. Food was scarce and, when and if we found it, we needed to eat it there and then.

I couldn't believe this was Europe. We had had no idea what it was going to look like but when we saw refugees, their clothes and faces totally black with filth, openly trying to climb on top of or under trucks, our hearts sank. We'd all been so pleased with our smart new clothes and the trendy sunglasses perched on top of our heads. How silly and naive we'd allowed ourselves to be for those two days in Athens. We really regretted leaving the UN camp. Watching my fellow refugees jumping onto trucks like hungry monkeys, I lost all hope. That night, we slept on the concrete floor of an abandoned, partly constructed building. The others told us their stories.

"I've been here for a year now."

"This place is crawling with agents. Every one of them a liar."

"Trust no one."

"The police have sticks. They will kick and beat you."

"The police put my friend in the hospital."

"We are on the shores of Italy. So much for Europe."

It was a cold, miserable night. The building was opposite a wire fence, which separated the dock entrance from the road. That specific area was referred to by the refugees simply as "the Fence." All night long we could hear the shouts and screams of refugees being caught and thrown out of trucks.

In Athens, Ahmad and Engineer had been tipped off about a supposedly reliable local agent called Borat, who operated in Patras. It was easy to find him: his height and shaved head gave him a striking physical presence. We'd been told that for a price he could put us in a truck and get us to Italy. He controlled a section of parking where the trucks waited to embark, and this was when he got his people on board. He claimed there was little security, and what guards there were didn't pay much attention anyway.

"It's a bit chaotic, but I get the job done," he said proudly. "Some people try for six months or more to cross, but not with me."

He said he could offer us the best deal in the area: "Five hundred euros each. I can accept an okay from Kabul or Europe." What he meant was something that I'd learned already about how the system worked: "Your family will keep the money with a third party acceptable to both sides. Only once you have passed through safely will the money come to me. If I fail you, your family doesn't pay anything."

None of us had that much money and, besides, I had no way of contacting my family, but we didn't let him know that. We told him we'd think about it. As we prepared to leave and find somewhere to sleep for the night, Borat looked at me. "You. I think I know you from somewhere," he said.

"I don't think so," I replied, fairly confident I would have remembered this man had I seen him before.

"Where are you from? You are very young to be making this journey."

"Nangarhar," I said, naming my province.

"Wait a minute. I met a guy from there. In fact, that's who I was thinking of. What was his name? You look like him. His name was Hazrat."

"Hazrat is my brother." This I couldn't believe. "Where did you see him? When? Was he well? Where is he?"

"Whoa, whoa," Borat said, laughing at my enthusiasm. "Slow down for a moment." He told me Hazrat had stayed with him for several months, about half a year previously.

"Are you certain it was my brother?"

"I am certain as I can be," he said. "He had a picture of you, and I see the resemblance. I sent him to Italy. He was traveling with a friend of mine called Hodja. They got as far as Calais. I don't know where Hazrat is now, but Hodja now lives in Rome."

Once he knew I was Hazrat's brother, everything changed. "For the sake of your brother, I will help you. Get some rest. Come to fence area B in the morning and I will show you how it's done. Then maybe you will trust me and we can do business."

THE NEXT MORNING, I WATCHED AS PEOPLE CONGREGATED around the Fence. It had been explained to us that each agent controlled a certain section. They were running everywhere, even running after trucks that had stopped for gas, climbing and clamoring underneath. I couldn't believe my eyes. I was sure the truck would crush them.

Borat waved us—me, Hamid, Engineer, and Ahmad—over to where he stood with a group of other worried-looking refugees. "Okay, so today I will show you so you understand my methods. Tomorrow we'll do proper business together and soon after that, I will ensure that you will make Italy. It should only take a few days for you to learn this."

He indicated to me, "Come, Hazrat's brother." He took me through a gap in the Fence to where all the trucks were parked. He opened the back door of a truck, which was full of furniture, and put me and three other men inside, then shut the door. Before long, it started moving. Almost as soon as it did, it stopped again.

The police opened the door.

"Get out."

The other guys ran; I followed. The police chased us. I didn't know how to get out of the fenced area so I just ran from side to side trying to escape the police. Hamid was still standing at the fence edge by the gap we had entered. He screamed for me to run toward him.

Borat was waiting to greet me. "This is normal. Don't expect to be in Italy the first try. It's going to take a few days. But now you know how it's done."

"I'm not doing that again. It's too dangerous."

He laughed. "Okay, if you don't want to go inside, try the other way."

A truck had just parked outside the perimeter fence at the gas station, and the driver had gone inside. Borat pulled me over to it and pointed underneath. "There's a small space near the tires. Go under and find it."

"No way. I don't want to die. I'm not doing it."

"Your choice." Before I knew it, another refugee had dived under the truck and was heading for the spot I'd just refused.

Borat looked at me disappointedly. "Your brother was braver, Gulwali. If you don't practice, you don't learn, and you'll never make it."

A few minutes later he called me again. I was getting tired already, and was really sick of this. He took me to another truck and pointed to the gap between the driver's cabin and the trailer. There was a piece of metal connecting the two parts.

"Okay, try this way. Sit on top of that."

I looked at him in disgust. But his words about my brother being braver had stung me—the Pashtun pride runs deep. I squeezed myself into the space. The metal was unbearably hot, so I put my bag underneath my legs to keep from burning them.

I think there were other people in the trailer. Police officers or soldiers were shouting. I heard strains of Pashtu and doors slamming. No one came near my hiding place.

The truck lurched forward. Hot engine fumes hurt my throat and eyes. As we moved, the tire noise changed, and I could hear metal plates clanking beneath the wheels of the truck.

The engine stopped and I could hear European voices.

I was on the boat. Borat had told me I could get out and rest at this point. But, despite a waft of the tempting smell of hot food, I didn't dare. Instead, I tried to sleep. At least my hiding place got cooler with every passing minute.

Two hours later, the truck drove me into Italy. I couldn't believe I had made it. The stories I'd heard had been so discouraging, but here I was—all thanks to Borat, who was as good as his word. I was sad to think I'd left all my friends behind, but what choice had I had?

We were moving quite fast now; the engine was unbearably hot. I took a rock from my bag, which was now starting to smell scorched, and banged it hard against the engine casing, again and again. The truck slowed to a stop.

I scrambled out of my hiding place just as the driver climbed from the cab. The look of shock on his face at being confronted by a filthy child said it all.

I was dazed, and couldn't think clearly—perhaps it was the heat and fumes. Instead of running, I just wandered the verge, staring at the distant farmhouses, rolling pastures, and neat rows of grapes.

Soon the police arrived, and led me to their car. They took me to a police station. They were gentle with me and treated me with decency. They were able to ask me some basic questions: What was my name? Where was I from?

They knew I had come from Greece, but had no proof of this. Otherwise they would have deported me back.

One of the officers made eating gestures. I nodded. We got back into the car and drove near the port. I thought they were going to send me back on a return ferry but, a moment later, we pulled up outside a cafe, where they bought me biscuits, croissants, and milk.

CHAPTER 24

HAPPY THAT I'D EATEN MY FILL IN THE CAFE, THE KINDLY Italian policemen gestured to me to get back in the car. I was relieved to see we were driving away from the port.

Before long, the sprawling maritime outskirts gave way to a recently built industrial estate. There were gray warehouses and the biggest supermarket I had ever seen.

I had seen plenty of deserts before—sparse, empty landscapes—but this felt altogether different. It was filled with signs of civilization, yet devoid of life at the same time.

It made me feel lonely.

Finally, the police car stopped outside yet another featureless building. We were met inside it by two women. One was slightly older. She smiled and pointed to her chest. "I am Sabine." The other one was a bit younger, and greeted me with a broad smile. She was pretty, with dark hair tied back in a ponytail. After my childhood of seeing women cover their heads, I still found the way European women left their hair free a little strange, but it no longer made me angry.

"Ciao, Gulwali," she said, holding out her hand. "Alexandria."

I smiled and pressed my palm to hers.

She was the first woman I ever shook hands with.

Sabine took me into a large beige room with comfortable brown sofas and a television in the corner, which was showing some kind of strange-sounding game show. She smiled, sweeping her arms in high arcs to make me feel at home. She was very expressive, using gestures as we couldn't speak to one another.

"*Thank you,* Sabine." I tried to communicate using the few words of English I knew by now.

Smiling once more, she ushered me down the broad hall. It was lined with doors, and light flooded in from a skylight above.

High-pitched laughter spilled out of a doorway the moment a door swung open. It revealed a slim, blond girl just a few years older than me.

"Katia," Sabine said. "Gulwali." She looked at us both. "Katia, Russki. Gulwali, Afghanistan."

I held a hand up.

A darker, fuller face appeared at the door.

"Ciao, Gabriella. *Permette che mi presenti Gulwali.*"

Suddenly I didn't want to shake this girl's hand. I didn't want to be introduced to anyone. I was depressed at being alone and having lost all my travel companions. It was still Ramadan and I wanted to be fasting. In fact, I wished I was back in the prison in Greece. I just didn't want to be here without any of my friends.

The two girls ducked back inside laughing, and shut the door. Hard.

Sabine smiled and gestured to me to come.

We walked to the end of the corridor, and into a bathroom. More gesturing from Sabine as she directed my attention to the toilet and showers.

Doubling back into the hall, we entered a bedroom—I assumed much like the one the girls had slammed the door on. It was small, also beige, with a window looking on to a car park. There was a single bed with two white towels folded on the end. A wardrobe stood in the corner and there was a chair next to it with a neat pile of washcloth, toothbrush, toothpaste, soap, and deodorant.

"Okay?" she said.

I nodded, and she began backing out of the room, pausing only to mime the option of having the door open or closed.

I returned in kind—closed.

When it shut with a secure clunk, I looked around the room once more. The wardrobe contained nothing but a few plastic clothes hangers. It didn't matter to me, because I only had the clothes I wore and a couple of T-shirts in my backpack.

I picked up the soap and sniffed it. It smelled sickly sweet. Throwing the towels onto the chair, I collapsed onto the bed and fell into a depressed sleep.

I AWOKE SHARPLY AND VERY DISORIENTATED IN THE HALF-light. It took a few moments to place myself. I guessed it was early evening.

The hallway was deserted, but I could hear the television playing loudly. A cold floor could not spoil the pleasure of the clean, lockable toilet and shower. Although I had managed to wash my face in the police station, I was still covered in grease from the truck. The hot shower felt wonderful, washing away the grime, but it did not wash away my distress.

I was very hungry, but was anxious about the common room. Making a cautious entrance, I found Katia and two teenage girls sprawled over the furniture. They were wearing tight jeans and cropped tops.

They barely noticed me. They were transfixed by a group of young men dancing and singing loudly across the television screen. The music sounded alien to me, and I felt oddly dehumanized, the way I entered and exited the room as if invisible.

I could smell food and hoped perhaps I might find Sabine or Alexandria.

My nose led me to an open-plan kitchen and dining room. A silver-haired man sat at a large wooden table, reading a newspaper. He maneuvered out of his seat and thrust a heavy palm at me.

"Ciao, Gulwali. I am Davide."

Ushering me to the table, he returned with a plate of pasta, chicken, and salad. I nodded my thanks, then stared at the chicken. Using my few words of English, I tried to explain I couldn't eat it if it wasn't *halal*. "I am Muslim."

He understood, and gestured to me to only eat the pasta and salad.

After I'd scoffed that down, Davide took the plate and returned with a sliced orange and a bowl of vanilla ice cream. That disappeared in seconds too.

When I had finished eating, Davide tried to engage me in conversation. Obviously I couldn't speak Italian, but he persisted, trying in English. But, really, I only had a couple of words so I couldn't make out anything he was saying.

We sat and stared at each other for a while. The music from the television continued to play in the other room. My stomach was full and heavy, and pretty soon its sad weight dragged down my eyelids.

I WOKE UP IN MY ROOM, SCREAMING. ALEXANDRIA FOUND ME on the floor. I was in a heap, crying and shaking. "Gulwali. Oh no!"

She switched on the light, but I couldn't bring myself to look at her. I got back into my bed and turned my back toward her.

I could feel her standing there watching me for a few minutes, before she was satisfied my breathing was calm again. She switched off the light and quietly left the room.

The next day Sabine—I think she was the manager—brought me some new clothes: new jeans, a T-shirt, several pairs of underwear, and a pair of trainers two sizes too large for me. I put the clothes on, but chose to stick with my boots from Athens instead. They also brought me pajamas. I thought they were a bit strange—the idea of a separate suit of clothing just to sleep in didn't really make sense to me.

I was safe, clean, clothed, and well fed. These people were clearly trying to help me, but I was confused and anxious. No one could explain to me what was happening or how long I would be there. And I couldn't stop worrying about Hazrat. Since learning he was alive, all I could think about was finding Hodja, the man in Rome who Borat had told me had traveled with my brother. If I could find this man, he might lead me to Hazrat.

Because of the language barrier, I couldn't explain any of this, so what was supposed to be a kind of sanctuary actually felt like a prison to me. The sad truth is, in my mind, the lack of information didn't make it feel that different than being with the smugglers—even though I know they truly tried their best for me.

Alexandria was in the living room talking to two teenage boys when I burst in and fired a string of questions at her in Pashtu: "What am I doing here? How long will I be here? Are you going to deport me? I want to go to Rome. I need to find my brother."

Alexandria just blinked at me, stunned by the ferocious onslaught of Pashtu. She must have registered my anxiety and frustration, which was quickly turning to anger. The two boys smirked, however, barely trying to hide their amusement.

She gestured to them to leave the room. They laughed loudly all the way down the hall.

I could feel tears building behind my eyes. I was a coiled spring of confusion—I knew this was a safe place, but I couldn't relax. I didn't want to be there.

Alexandria beamed her kind smile and gestured to me to sit. She switched the television to Al Jazeera Arabic, and signaled that I should watch and wait, while she made a phone call. I think she thought I understood Arabic, but aside from knowing the Quran in Arabic I didn't. But I watched the pictures.

She returned and leaned over me. "Gulwali? Farsi?"

I smiled. Did this mean they were bringing someone to talk to me? "Pashtu?" I looked up at her hopefully.

She shook her head.

I tried again. "Dari?" I pointed to my chest. "Gulwali. Pashtu. Dari." I was trying to say to her that I spoke Pashtu and Dari.

She smiled ruefully and shook her head again. "No. Farsi."

I nodded, then shook my hand from side to side as if to say, "Okay, a little bit."

She smiled and left the room. She soon returned, making steering-wheel signs. Time for a drive.

"Where are you taking me?"

Alexandria just smiled. I could feel my frustration growing again.

A short time later, I was sitting in another beige-colored office in a local government building, with Sabine the manager by my side. Across the desk from us sat a stern-looking woman in a tailored suit. To me she looked so smart and important I thought she might be the mayor. She and Sabine were talking in bouncy, rhythmic Italian. "I am Fabiana," she said to me. "Hello, Gulwali."

She spoke English, so I could make out a little bit of what she said. I couldn't help but like her, but that didn't change my feeling of mounting fear and anger. Physically, I was better off than I had been in months; emotionally, I was imploding.

Fabiana picked up her telephone receiver and dialed a number. She turned on the speaker phone, so we could all hear it ringing.

More Italian followed, before the voice on the line switched to Farsi. "*Salaam,* Gulwali."

I sat upright, jolted with surprise.

In my best Farsi I said hello back. I told him I was a Pashtu speaker but I could understand some Farsi. But I asked that if he couldn't speak my language of Pashtu, could he try speaking to me in Dari—the Afghan version of Farsi?

The speaker replied that he spoke zero Pashtu, but suggested that I should try to explain myself in Dari and he would respond in Farsi.

Throughout the journey and my time in Iran, my Farsi pronunciation had improved greatly, but on the road I'd gotten used to the different smugglers and agents speaking a bastardized version of Farsi mixed with Kurdish. I was mixing all my languages together. In a gabble of blurted-out Dari, Farsi, and Kurdish in my native Pashtu accent, I tried to tell him everything. That I had lost my brother and was trying to find him; that I wanted to know how long I had to stay in the home; that I wanted to know what was going to happen to me; and that I wanted to know how long I had been on the road already.

Unsurprisingly, he struggled to understand a word. "Calm down. Speak slowly. Calm."

The whole thing was made even more frustrating because his Farsi was spoken with such a strong Italian accent, I couldn't really follow him either.

My head was bursting with questions and I desperately needed answers to stem the turmoil inside, but there was no way I was going to get them. I persisted for a while, but it soon became apparent we were making little headway.

When the call began, Sabine and Fabiana had looked expectantly hopeful; I think the dejected look on my face told them the real story. After I handed the receiver back to them, the man on the other end spoke to them in Italian, probably confirming we hadn't achieved much.

"Sorry, Gulwali," said Fabiana. I felt bad for them. I know they really wanted to help, but if we couldn't communicate, how could they explain the processes to me?

When we got back to the children's home, Davide was busy in the kitchen. The other children were sitting around the dining table. They stared as I joined them.

Davide placed a large platter of spaghetti and meat sauce in the middle of the table. He returned with a basket of bread, and put a brimming bowl of pasta and vegetables in front of me.

"*Halal,*" he said triumphantly.

The three girls who had ignored me in the common room chatted in Italian. The Russian girl just stared at her plate. I ate in silence. That night I settled down in my little room. Despite my exhaustion, I had trouble going to sleep. I had suffered fear, hunger, thirst, brutality, even cruelty through my travels, but I had never been alone. I had always been in a group of men and always had someone looking out for me. At home I had shared my grandparents' room.

This was the first time I had ever had a room to myself. I didn't like the feeling.

EVERY DAY CONTINUED PRETTY MUCH IN THE SAME WAY. I paced the floors like a caged animal. I'd been in a constant state of adrenaline, always on the move, for so long that I couldn't cope with being still or not allowed outside. All the trauma and pain of leaving my family flooded back and I couldn't stop the memories from overwhelming me. Every night I awoke to the same nightmares.

It felt as though I had been there for ages. I started to get disruptive. They took me to see the lady in the smart suit, Fabiana, at her office a second time—something that made me even more furious, because it was patently obvious that although she was trying to reassure me, it wasn't working.

The person to whom I felt the most connected was Alexandria. I used to stand on the third-floor balcony looking down at the car park, waiting for her to come to work. Whenever I saw her, I felt some measure of relief and comfort, but usually this came out in angry emotion. She bore the brunt of my frustration.

"I have been here one month," I managed to say in English. "One month. Why?"

She shook her head patiently and gestured me into the kitchen. There she poured me a glass of juice and sat me down. She took the calendar off the wall and pulled up a chair next to mine.

"Gulwali. No. Ten days you have been here. Ten. Only ten." She gestured with ten fingers, then pointed at the date on the calendar. "See? You came this day. Today is this day."

I looked at her in disbelief. Could it really have only been ten days?

The following afternoon, Sabine ushered us all into a waiting minivan. The others seemed really happy and excited, but continued to ignore me. Sabine tried her usual mime routine, this time making her hands rise and fall, accompanied by spluttering sounds. Then she flapped her hand through the air, screeching so that the others all burst into fits of laughter.

I smiled and nodded as though I understood.

As we parked on the seafront, I worked out that Sabine had been trying to act out a beach scene for me. We had come to a nearby seaside town.

The beach was rocky, with large boulders. It was early October, and the wind was picking up. The waves were large. I froze with fear as flashbacks from the boat came flooding back. I didn't want to be anywhere near the water. The others started running along the beach, chasing each other. All I remember is that I wanted to get away from there. I wanted to run. To get as far away as possible.

Later, up on the promenade, near some shops, Alexandria stopped me outside a clothing shop, and held a red sweatshirt to my chest.

"*Bella?*" she said, deliberately inflecting the word to convey her question.

I nodded.

The two Italian boys stood some distance away, laughing and aping my every movement. I shot them a self-conscious glare.

There were a lot of people walking around, with dogs on leads. I also looked at the pavement cafes, with people sitting outside drinking coffee. In the central square, teenagers were rushing past us on skateboards. I'd never seen these before and fully expected one of them to hit me or slice my feet off with it.

It should have been a pleasant, interesting afternoon out, but for me it was a trip filled with terror, confusion, and loneliness.

I started plotting. I had decided I was going to run away.

Sabine returned with the girls and ice creams. Mine was pistachio—it was delicious. In Jalalabad, the city near my childhood home, my grandfather used to take us to an ice-cream parlor where they made the best-ever homemade ice cream. The taste of this took me right back there, and to being six years old again and happy.

We walked back down to the beach, to where the minivan was parked.

I hung back from the others a little. A bus had caught my eye. I thought about trying to climb aboard. The driver was standing outside having a cigarette and chatting with some local people.

I could have easily snuck on unnoticed. I had, thanks to the agent returning to me my family's 300 euros in Greece, more money than I'd ever had on me before.

I was torn. As desperate as I was to get away, I didn't really have a plan, and nor did I want to get Alexandria, Sabine, or the other staff in trouble. If I ran, surely they'd be blamed.

We went home and I took a shower. Alexandria had made it her mission to get me washing every day. At first I had thought she was crazy: in an Afghan winter I might have washed my entire body once a week. Our winters were so cold none of the children wanted to wash much; it also took my mother and aunties a long time to heat the water over the fire. In summer, I would swim and bathe every few days in one of the many rivers that flowed through our district. To me, showering every day seemed like a huge waste of water. "But I'm clean," I protested to Alexandria in my best English. She would just shake her head and point at the bathroom door. I liked her so much that at these little moments, she did find a way to burst some of that boiling anger inside me. I think she may even have persuaded me to laughter.

Her persistence was such that I almost began to enjoy the daily ritual. And, as I grew more mentally withdrawn and trapped, the shower also became a good place to cry. I felt like tearing the world apart, just so it would know how I felt.

ANOTHER FEW DAYS BLENDED INTO OTHERS. I HAD A SENSE OF time based on when I had last fasted: it had been Ramadan in Greece, and that meant that Eid, the celebration that comes after a month of fasting and is a bit like the Muslim Christmas, was just around the corner.

As a child, Eid had been something I looked forward to all year. We got new clothes and money, as presents, and the whole family would dress up and go out to picnic sites for the day. Eid celebrations had been some of the happiest and best moments in my childhood. I called to mind bittersweet snapshots of playing egg fights—with my brothers. You bash each other's boiled egg and whoever breaks another's shell gets to eat it. Time and again Noor and Hazrat had a knack for finding the strongest, hardest shells and beating me. I remember glaring at them as they shamelessly stuffed my eggs into their mouths after winning. One time I had been so angry at Hazrat I thumped him, and my father had told us all off. I had detested my brother then. Now all I could think about was finding him.

I ATE BREAKFAST AS NORMAL, BUT INSIDE MY HEART WAS RACing. I was preparing myself. I was nervous, but I told myself if I could run away from the Iranian police, I could easily escape a children's home.

Back in my room I carefully packed my new clothes into my bag, along with the toiletries. Then I pushed my window as far open as it would go, and climbed onto a ledge that ran beneath it.

I had to do this. Anything to stop the feeling of being caged.

I took a big gulp, said a prayer, and dropped three stories to the car park.

I landed with a thump. The adrenaline was surging so much I didn't feel any pain. I looked up at my window, a little shocked at what I had just done. After a few seconds, however, the pain kicked in: the back of my heel was throbbing. But there was no time—I had to get away.

I exited the car park—the road ahead had three different junctions. I had no idea where I was going other than to try to find a bus or train station. I ran blindly down one of the roads.

Once I had gone some distance from the home, I slowed down and tried to act casually. I managed to make myself understood by using a bit of English: "Excuse me. Where is station?"

An old man directed me to the left.

I saw train tracks and knew I was headed the right way. Escape was imminent. Crossing the roundabout at some traffic lights, I saw the station building just ahead.

Just then Alexandria screeched up in a car, screaming from the window: "Gulwali. What are you doing?" She leaped out of the car and grabbed me by the shoulders, shouting at me in a stream of Italian. She was shaking as tears streamed down her cheeks, and I could see that she was as worried about me as she was angry. She kept pressing my shoulders and arms to feel if I was injured.

I had picked up a bit of Italian in the fifteen days I had been there, so I had a good sense now of her flurry of urgent words. I was a boy, I had nowhere to go. What would I have done?

I felt really guilty, but I still tried to resist getting in her car. "No. I am leaving."

At this, she got really angry, making a telephone-to-the-ear gesture. "*Polizia,* Gulwali. I will call the police."

I got in the car. She was still crying, and by now her tears set me off too. We both cried all the way back to the center. She was shaking so much she lost grip of the steering wheel and we almost had an accident, bumping into another car.

When we walked back into the building, it was into a sea of shocked, ashen faces. I felt like I was walking into the lions' den. They were all judging me, but none of them understood my pain.

I went back to my room where not even Davide's cooking could lure me out. I lay on my bed trying to calm myself down, but my anger wouldn't subside. I felt humiliated. If anything, I was even more determined to leave. I could hear Alexandria on the phone. Her voice was nearly hysterical. I guessed she was either talking to Sabine or Fabiana about my attempted escape. I felt really guilty for putting her through that, but I also feared she wouldn't like me after this. Now I'd probably lost my only friend in this place too.

I *had* to go.

I came out of my room and into the main sitting room. Courage swelled inside me like a balloon. The other kids were in there, watching TV.

"I am going," I shouted in Pashtu. The girls all looked at me like I was crazed. The two boys looked a little scared, which pleased me. "I am leaving. Good-bye."

So fueled was I by a sense of righteous anger, I couldn't resist a final flourish in Italian for their benefit. "Ciao."

I unlatched the main door and stormed out.

I was in the car park when I heard Alexandria's voice shouting from the balcony above: "Gulwali. Gulwali. Stop."

I looked up at her briefly, imploring. I wanted her to let me go. In my head, I willed her not to come and look for me.

I knew she'd be following any second, so I broke into a sprint.

This time I knew where the train station was.

I think I had expected to see cars full of police waiting to arrest me, this escaped child convict. But the station was quiet except for a couple of pensioners and a woman with two young children.

Breathlessly, I spoke to the man at the ticket desk, using a mixture of my limited English with the few words of Italian I had picked up in the home: "Rome, please. I wish to go Rome." It was the place I knew Hazrat had been—I could start there.

He printed me a timetable and, as I tried to calm my panicked breathing, he managed to explain to me that I needed to get a connection to a town called Bari, from where I could get a direct train to Rome.

He looked at me curiously as I handed him a 50-euro note for what was a 5-euro ticket to Bari. I feared he might try to stop me.

The train pulled into the platform. I ran and jumped on.

ON THE TRAIN I SANK INTO THE SEAT. AS THE ITALIAN LANDscape sped past the window I was relieved to be free, but in the back of my mind I was worried about myself. I kind of knew that this was a silly thing to do—these people had been caring for me, so why couldn't I accept that? Why was I running away from the first genuinely safe place I had known? I felt oddly discombobulated, almost like a feral version of myself, trusting no one.

I think if they had outlined what had been happening to me, and what would happen to me, it would have been easier.

It didn't take too long to get to Bari. Once there, I managed to buy myself a ticket to Rome, which cost me 40 euros. I also bought myself a sandwich, some chocolate, and a drink. This little journey had cost me 50 euros so far, leaving me with 250.

There were a couple of hours to wait before the train for Rome departed. I cowered on a bench on the platform, expecting at any moment the police to swarm around me.

I REGRETTED ROME ALMOST THE MINUTE I ARRIVED. IT WAS late, and darkness had fallen. The station was huge and scary. People flowed back and forth in all directions; I so desperately wanted to ask someone for help, but I didn't dare.

I had been told in Greece that in Rome I needed to find the park, where all the refugees stayed. As I left the station I could see a few refugees—I think Eastern Europeans, but I wasn't sure—begging by the side of the road. I walked closer toward them and noticed a group of Afghan-looking men walking down a side street. I followed them. They were stopping passersby and showing them a piece of paper, asking for directions to whatever was written down. I had a hunch they were looking for the park too. A lady directed them to a certain bus stop. I followed. When they got on the bus, I did too.

This was my first time on a bus in Italy so I had no idea how to pay for a ticket. I had money, so buying it wasn't an issue, it was just the how. The Italian people seemed to have tickets already—they were pushing them into a slot in a machine by the door. The machine sucked the ticket in then spat it back out.

The other Afghans just walked through the carriage and stood at the back. The driver didn't seem to be checking, so I did the same.

Ten minutes later, they got off the bus by a vast park and I followed, unnoticed. The scene was incredible.

The park was surrounded by beautiful, historic buildings that made me gasp in awe, but everywhere else were bodies. There were people lying on cardboard cartons, on the footpath, on the side of the road, by the carved marble monuments, in the park grounds, under the trees—it was a shantytown in the middle of Rome.

I had that realization again that half the faces in the world were represented there. I recognized Ethiopians, Eritreans, Sudanese, Congolese, Somalis, Iraqis, and Afghans. There were probably many other nationalities I didn't recognize. But the overwhelming majority were Afghans or Iraqis.

I obviously looked a bit lost. "What's the matter, boy?" a narrow-faced man asked me in Pashtu.

It was such a relief to be able to understand what someone was saying to me, but I didn't feel I could reveal all to a total stranger. "I just arrived, *Kaka*."

"Be careful, boy. This place is dangerous and there are some bad men here. Stay away from the police. They come at night."

"Oh." I really didn't want to meet the police.

He pointed to a small side street. "There is an Afghan Internet cafe there, where you can call home. There are many food places where you can eat. To sleep, come back here."

"Thank you, *Kaka*."

"Where you going, boy?"

"England."

"Good. Keep moving. They don't like us here in Rome. You keep going, and don't stop until you make it."

As I followed his directions, I was increasingly furious with myself. Why had I left that nice, safe home? What had I been thinking? Poor Alexandria must be in big trouble. I wished I could see her and say sorry. A large part of me couldn't understand why I had run away. They were trying to look after me. Why couldn't I have accepted that? Why did I behave like a wild animal when all they had shown me was kindness? Then Hazrat's face flashed before me and I felt that overwhelming stillness again, the same calm I felt after prayer. I had to believe that fate was taking me somewhere else.

I found the nearby Internet cafe. There were little booths with pay phones in them, and the man behind the counter seemed to be arranging transportation, and offering train tickets too. He was obviously Afghan. He looked surprised to see a clean and well-dressed young Afghan walk in.

We started chatting.

"Do you live here?"

"No, I just arrived in Rome. I am hungry and I have no place to stay. What should I do?"

"Do you know anybody in Rome?"

I hesitated, then went for it. "Yes, my brother's friend is here. I don't have his contact details, but his name is Hodja and he is from Nangarhar."

I couldn't believe my ears at his next words.

"I know Hodja. If it's the same guy, then he's a good friend of mine."

He told me his name was Marouf. He gestured to the phones. "Do you want to call your family?"

I shook my head. I still didn't have any contact details for mine, but I didn't want to tell him this. "I'll do it later."

He was busy with customers, so suggested I go and eat somewhere. He promised that in the meantime, he'd try to contact Hodja for me.

I went to a Turkish cafe. I couldn't read the menu above the counter so I signed to the guy to make me something nice. He returned with a can of Coke and a lovely hot sandwich with egg, chicken, chips, and cheese all mixed together. It was delicious.

When I went back to the Internet cafe, my new friend was smiling. He'd already spoken to Hodja but got him back on the phone so I could talk to him and ensure it was indeed the right man.

"*Salaam*. Are you Hodja?"

"Yes. Are you Gulwali? Are you the son of the doctor?"

As soon as I heard this I was so happy. I knew it was the right guy.

We chatted some more. He assured me that Hazrat was well but that he'd tell me everything he knew when we met face to face. He had already asked Marouf to give me a bed for the night, and would see me tomorrow.

I thanked him profusely.

By now it was very late. Marouf told me to go back to the park and come back when the shop was closing, around midnight.

As I walked around the park in the cold, late October air, I looked about me at the human misery. I was relieved to not be sleeping there, but worried about staying with a stranger.

When I got back to the shop it was locked and in darkness. I banged on the door.

Marouf opened it. "Shush."

I felt scared—the situation didn't feel at all comfortable. I'd only met this man two hours ago.

I had assumed he had a flat above the shop or some rooms at the rear. Instead, he pointed to a small foam mattress behind the counter.

"We need to sleep here tonight."

I felt like running. Why would this man help out a young boy? Suspicions bloomed in my mind.

"Apologies, Gulwali. My house is far from here. Tomorrow is Eid. We'll go to the mosque together for Eid prayer, and Hodja will meet us there."

I was only slightly reassured. It still felt odd to me he was letting me sleep there, and not with the hundreds of other desperate refugees in the cold park.

I knew some men took boys like myself as *bachas*—I'd already been warned about this. *Bachas,* traditionally, are Afghan dancing boys, although they offered a far wider range of entertainments for their masters, very little of which was consenting.

I told myself if he was Hodja's friend, maybe this was really just a stroke of luck.

I pulled a sweater from my bag for a pillow and tried to sleep, which wasn't easy, given how worried I was. Trying to calm myself, I realized that I had known Eid was soon but had had no idea it was as soon as tomorrow. I tried to feel happy about that at least as I dozed off.

IT WAS STILL DARK WHEN WE WOKE UP. WE WENT TO A nearby supermarket, which my host said was owned by a friend of his. There we showered and had breakfast. My host had a bag containing his new Eid clothes: a lovely, embroidered, proper Afghan *shalwar kameez.*

This made me even sadder. It was my first Eid away from my family: *Eid al-Fitr*—the Islamic holiday that comes after a month of fasting.

That thought came with a jolt of realization that I had been on the road for a whole year. That meant I was thirteen years old now. I had completely forgotten about my birthday—I realized I had unwittingly marked my passage into teenhood by jumping from the youth center window in Italy barely twenty-four hours before.

I didn't really care about my birthday—it was, after all, just another date—but it did make me feel more lonely and isolated. "Toughen up," I told myself. "You are a man now."

Marouf took me to Rome's central mosque. It was magical—the largest mosque I had seen outside Istanbul. It was raining, but that didn't dim the Eid atmosphere: children in new clothes ran around outside, and there were food stalls selling all manner of deliciously scented kebabs, sweets, and rice.

Inside the mosque I was thrilled to see the diversity of Muslims there—all colors, races, and ages. I felt a strong sense of unity and brotherhood, something which eased the pain of not being with my family.

After prayers, we walked outside again, through some trees and flower beds. Marouf walked over to a bench and introduced me to a bearded man. I was taken aback—Hodja reminded me so much of Uncle Lala.

He embraced me. "Welcome, Gulwali. Your brother told me so much about you, young man." He peered at me. "But you don't look like the little kid your brother said."

To me, being in Rome, being at the mosque, the hug, that fact that he knew my brother—the whole thing felt like an Eid miracle.

"Let's eat, and then we'll talk."

After we'd had our fill from the amazing food stalls, which offered an array of edibles that matched the nationalities of those inside—African, Arabic, Asian—we sat back down to talk.

Finally, I heard the news I had been so desperate to hear.

Hodja told me that he had been with Hazrat a few months earlier. They had been traveling together from Turkey, to Calais in France. He had been looking out for Hazrat. "He was always concerned about you. He never stopped talking about you or worrying about what had happened to you. He asked every agent or smuggler if they had seen you or to look out for you. He had a passport-size photo of you that he showed to everyone he met. He was determined to find you." My heart sang to hear this. He went on to explain that after a few weeks in Calais, Hazrat had managed to get on a truck to England. "He was so brave. It's hell in Calais. I only stayed as long as I did because he kept urging me not to give up. After he succeeded, I kept on failing. It's a terrible place. In the end, I couldn't take it anymore so I gave up and came back to Italy."

"So Hazrat is definitely in England?" I asked.

"Yes. Some of the guys shared a mobile phone, in Calais. He called to tell us he had safely arrived, but since then I've had no way of contacting him."

I had mixed feelings on hearing all this. On one hand, I was very glad Hazrat was safe in Britain; on the other, I felt depressed that I was still en route, facing all this hardship and these challenges alone. I was really happy to know Hazrat had been as worried about me as I had him, if not even more so, but I wondered how, even if I did make it England, I'd ever find him.

What I was absolutely certain of was that England was where I had to go if I had any hope of seeing my brother again.

For the first time since jumping out of that window, I was glad I had run away from the children's home. I still felt really guilty about Alexandria, Sabine, and the other staff for letting them down, but I hoped if they knew my story they wouldn't be angry and might even understand why I'd run.

Marouf had to get back to his shop but made Hodja promise to bring me to say good-bye before I left. I thanked him from the bottom of my heart for letting me stay with him the night before. I felt guilty that I'd been so suspicious of his kindness.

After mosque, Hodja took me to his house, which he shared with two other Afghan men. I stayed there for two nights. Hodja had a daytime job doing some kind of manual labor, I don't know what. He insisted on buying me new clothes: a really warm, bright-yellow ski-type jacket and new jeans. He also went with me to the station to help me buy my train ticket to the French border. He explained that I needed to get a train from Rome to Genoa, then from there to Ventimiglia on the Italian/French border. He paid for my ticket that far, but explained that from the border I'd have to buy new tickets for the French railway in the nearest French city, which was Cannes. And from there, to Paris.

"In Paris, talk to some other Afghans and make a wise decision," he advised. "Calais is very dangerous. Please try not to go there alone."

As he handed me the tickets, he gave me 150 euros.

"No, I can't take this. You've done so much for me already. This is too much."

He insisted: "Gulwali, you will need this in Calais. I have been there. Trust me."

"No, I have money. I still have some dollars and two hundred euros. I have lots of money."

He laughed. "It might seem a lot to a child, but the train tickets in France will be more than one hundred euros. And there will be many people trying to part you from your money in Calais. You will need to survive there."

He went on to advise me what to do when I finally got to Calais.

The most sensible thing, Hodja said, would be to find a good and trustworthy agent—well, the most trustworthy I could. I was to give them my cash but make sure they understood that it was all I had and that no more would be forthcoming.

If I could convince them, Hodja finished, they might guarantee to take me more quickly. He said the smuggling operations in Britain and

France were highly organized, with networks everywhere. Where the system had gaps was that the smugglers on the ground in Calais were often short of cash, and someone waving a fresh 100-euro note in their face might get lucky.

He ended with a word of warning: "Handing over your cash is more risky. An untrustworthy agent might run away with it—I know people this happened to. This is why people prefer the usual system, which is guaranteed."

Guaranteed. That word again. In my experience, nothing on this journey was ever guaranteed.

The train wasn't for a couple of hours, so we went to say our goodbyes to Marouf at the Internet cafe. He was delighted to see us. But his next question threw me: "Gulwali, can you do something for me? I'd like you to travel with my friend, Shafique." He pointed to a teenage boy standing behind him, who looked as confused as I did.

"What?"

"Shafique needs to go to France. Travel together, and you will both be safer."

"No way. I'm not some kind of agent. Why do you take me for a smuggler?"

Hodja intervened. "Gulwali, no one is asking you to smuggle him. It's a good idea. You shouldn't be alone, and neither should he."

I wasn't happy—I could barely look after myself, let alone someone else. But Shafique looked so worried, my heart went out to him. People like Baryalai had looked out for me, so why couldn't I help someone else, now that I had been asked? I also felt that I owed Hodja and Marouf for their kindness.

So I agreed.

LATER THAT NIGHT, SHAFIQUE AND I BOARDED A TRAIN TO Genoa. There were a lot of Afghans on board, many of whom insisted on repeatedly asking the conductor how much longer it was until we got to Genoa. I told them to stop asking him. I was really worried it looked suspicious, and that we would end up getting detained.

On the train, Shafique and I got to know each other a bit better. He was sixteen and from Kabul. His reasons for running away were not too dissimilar from my own. His brother had been killed while fighting with the Taliban against the Northern Alliance. He had been in Italy for a month, sleeping in the park with the other refugees. He had attempted to get to France once already, but had been sold the wrong ticket and ended up in Milan.

If you look like an illegal refugee, he explained, the Italian train staff normally refuse to sell you a ticket to the border unless you can show ID or a passport. For most people arriving in Rome, it's the first time they are without an agent. Like my family, most families only pay for the trafficking up to Greece, so they are very vulnerable and have no idea how to even buy a ticket. Shafique told me that selling tickets to refugees and migrants was big business in Rome. He had been tricked into handing over money for what he was told was a discounted ticket to France when in fact it was to Milan. After that, he'd somehow managed to get back to Rome. The day before, he'd gone to the Internet cafe to buy a ticket from Marouf—dodgy tickets being another of his many enterprises. Marouf had told him about me and suggested we go together.

After my initial displeasure, I was actually quite pleased to have a new friend. Once again, I was grateful for the stroke of fate that had brought me to him.

When we pulled into Genoa station, late at night, Shafique and I ran from the train. The other troublesome Afghans had fallen asleep. There was no time to wake them up.

The station was much bigger than I expected; we had no idea how to find the connecting train to Ventimiglia. We stared at the flashing words on the departure board but could make no sense of them. I took a deep breath and approached a conductor, who took the time to read my ticket and point to the right platform. He looked at his watch and urged us to hurry.

I smiled and waved, and we ran. We couldn't find the way he'd described but we could see the platform on the other side of the tracks. There was no time. I then did possibly one of the stupidest things I've ever done: I leaped onto the tracks and ran across. "Shafique. Come on."

It was late so there weren't that many people around, but those who were there watched in disbelief. It was idiotic of me, because it was a sure way to attract attention and get arrested.

Shafique followed me, but when we scrambled up the other side to the platform he was furious. "What did you do that for?"

The worst thing was that we realized we were on the wrong platform. We ran to the end, then around a corner, where we spotted a train driver walking toward a train. I ran over and showed him my ticket. He smiled and nodded for us to follow him.

In the morning, we arrived in Ventimiglia. Heading to the departure board to try to work out how to get the train to Cannes, we were deliberately trying to walk normally and casually, but still an Italian police officer approached Shafique and me. He asked to check our ticket. When he had done so, he shook his finger at us. "No France," he said in broken English. "No France. Go back or go outside."

As we reluctantly walked outside, I looked back and realized how we'd been caught. Less than twenty meters behind us were five Afghan men. They had been following us without us knowing.

"Why did you walk so close to us?" I was so angry. "You got us caught."

The policeman was walking behind them, escorting them out. He walked back and stood at the entrance to the platform where the train was leaving for Cannes and France. There was no way we were going to get past him.

The seven of us drifted around the streets until we found a cafe, where we bought some food. I was still pissed off with the other guys, not least because they now stuck to us like glue. Shafique and I were both kids, but they seemed to be completely useless at this.

As we ate, an argument broke out about train tickets. It was clear we had to risk crossing from Italy to France without passports, and while buying tickets within Italy wasn't such a problem, the moment we tried to cross a border we needed to show a valid passport or ID to buy a ticket. None of us had any. Added to that, none of the men wanted to go to the ticket office to try to buy tickets.

"I'll go," I said, confident of my ability to win the ticket salespeople over. Although Shafique was sixteen, three years older than I was, I had assumed the role of boss. I think because I had been on the journey longer than he had, I felt more experienced. But no way was I doing this for free: "I'll sort it out, but you all have to contribute to pay for my ticket."

I circled the station entrance, scouting for the policeman who had stopped me earlier. He wasn't there, although a number of other policemen were patrolling. I strolled past them, being sure to keep my head low. In my new clothes and bright-yellow ski jacket courtesy of Hodja, I was no more threatening than a sulky Italian pre-teen.

I approached a middle-aged lady sitting behind a glass screen and gave her a broad smile that I calculated was both charming and vulnerable. Women had been kind in the past, and I hoped this would prove to be true once more. My Italian vocabulary was still very limited, but I did have the few phrases I learned in the children's home. I also had a map of the rail network that Hodja had given me in Rome, which helped make the point.

After a lot of effort, I succeeded in making her understand that I wanted to know how much a ticket was to Cannes.

At first, she wasn't willing to even tell me that: "Sorry, ID please."

I tried to gesture and explain my friends were waiting for me. She wasn't happy, but I just continued to give her my best pleading face and, in the end, she wrote down the price for me.

I left the station and found the others.

"Here's the deal," I said. Their eyes were fixed on me. The simple act of buying tickets was very stressful for refugees, and I knew they couldn't do this without me. "It's fifteen euros each for your tickets, plus the cost of my ticket spread over the five of you—but not Shafique—that's three euros. Think of it as a booking fee. Total eighteen euros each."

There was some grumbling, but they had no choice.

I knew my little money-making scheme was immoral. I tried to justify it to myself, although it still troubled my conscience.

I went back to see the lady. I can't say she looked thrilled to see me.

I'm not sure if it was because I was persistent or persuasive—probably it was my youth and vulnerability tugging on her heartstrings—but she began to soften slightly. I got the feeling she would sell me a ticket, but the issue was the other six.

"Why seven?" She shook her head. "Too many people."

I just shrugged and continued to give her my best pleading face and, in the end, amazingly, she sold me the tickets. Although, if I understood her Italian correctly, she did give a stern warning that we were likely to be arrested.

She pushed the tickets across the counter, looking around a little nervously. It was then that I realized that she was probably risking her job to help me. I gave her a genuine smile. She looked scared, but her eyes spoke of her emotion. I was beginning to realize that there were kind people around who were truly moved by the plight of refugees.

I knew that a group of refugees was more likely to attract unwanted attention, so I took control and ordered everyone to board the train in pairs, and get on different carriages. My days of moving around the different *musafir khanna*s in Istanbul had taught me something.

When the conductor came into the carriage, he stopped and checked our tickets. I didn't like the way he looked at us, but he moved on quickly.

I panicked when two police officers walked through the carriage.

"Shafique, police."

We dropped our heads. Shafique pretended to sleep.

I stared at the Italian countryside sliding past, all the while carefully checking their reflection in the window.

They went past us without a second look.

Shafique raised his head and grinned.

An announcement then crackled out over the speaker system. I didn't really pay any attention—my Italian was so limited I only understood a few words—but this time it was different. "Shafique, I think he is talking French. I think we are in France."

We sat and listened to the rhythm of the address.

"I think I love the sound of this language," Shafique said.

We burst out with relieved laughter.

We weren't out of the woods yet, though. We listened for any announcements that mentioned our destination, Cannes, but neither of us could understand the accent.

I was puzzling what to do when we had a stroke of luck—a man entered our carriage. It was a young guy, with a rucksack slung over one shoulder. I didn't know him, but I could tell he was an Afghan Hazara.

"Hey," I said in Dari.

He turned. "*Salaam*."

"Do you speak Pashtu?" I ventured.

"Yeah. A bit."

We beckoned him to join us.

He sat opposite us, his bag in his lap. He told us he was a university student in Marseilles. He was generous with his advice, giving us a crash course on French pronunciation and useful phrases, telling us what time the train was expected in Cannes, and what to expect on the journey to Paris. As we slowed into Cannes, his final advice stuck in my mind: "Be careful of the police. They will give you a very bad time."

I wanted to know more, but we only had a few seconds before our stop.

"Will they put us in prison or deport us?" I said.

"Probably not—even if you want them to. Good luck."

That confused me.

"Gulwali." Shafique called from the train's open door. "Let's go."

The other guys had made it too. I half expected not to find them on the platform at Cannes, but I was happy they were there.

The station was full of police. We headed outside and found a grimy Turkish cafe near the back of the station.

"We met a guy on the train who said tickets to Paris are one hundred euros," I said.

The others blew out their cheeks at the price. An argument erupted, with everybody shouting over each other, trying to be heard.

"Hey. Heeeeey."

We all turned. The owner—a heavy Turkish man with impressive black caterpillars marching across his brows—leaned over the counter, waving a long kebab knife. "Enough of your noise. Keep it down or get out of my restaurant," he shouted in a mix of English and Kurdish. "And you can all buy some more food, too. I'm not running a charity. Bloody refugees," he muttered, turning his attention back to the rotating slab of grilling meat.

We lowered our heated bickering to a loud whisper.

Everybody would pool their finances, and I would buy the tickets again. I had hoped I could get another free ticket, but at 100 euros no one would agree to it.

We only had enough money for four tickets.

That was a major problem—but I had a plan. When the seven of us climbed aboard the train later that afternoon, three of the men went and hid in the toilets.

Shafique and I sat near the other two guys, trying to look innocent and unthreatening. A game of cat and mouse ensued with the conductor. We swapped seats and tickets on several occasions, the guys in the toilets trading places with those in the carriage.

I tried to look confident and relaxed when the conductor came past, knowing that although he had already checked my ticket, it was now in the hand of one of my traveling companions two carriages further along the train.

One of the guys got caught, though. As we pulled away from a rural station platform, there he was on the platform, looking horribly dejected, a policeman standing next to him.

It was a long journey.

CHAPTER 25

My gritty eyes slowly opened. We had been traveling for almost two days on this train since leaving Rome.

"Paris," Shafique gave a satisfied grin.

The train carriage rocked gently from side to side.

"Why is there so much writing on the walls?" I asked, as if Shafique might have an idea what the cartoonish images and words I saw through the carriage windows might mean. I had seen similar graffiti, although I didn't know what it was called, in the poorer parts of Greece, Turkey, and Italy, but I hadn't expected to see it in Paris.

For some reason, I had a notion that Paris was the poshest and most beautiful city in the whole world. I don't recall from whom I had heard this—maybe one of my teachers at school—but as a child in Afghanistan I'd been told that every morning an airplane sprayed a fine rain of pure perfume over the city so that the air always smelled fresh and beautiful.

As the train slowed into the station approach, it hit me that this was a lie. We pulled into the biggest station I had ever seen—a vast, steel-framed hangar, bustling with people.

As we made our way out of the station and onto the streets of Paris, we looked about us in wonder. The busy streets were lined with pale stone buildings with shiny shop fronts. There were restaurants and cafes everywhere, with smartly dressed people scurrying along so quickly it looked like a matter of life and death. Cars drove bumper to bumper as motorbikes buzzed past them like wasps.

"We need to find the park," I said, unsure of how we might actually achieve that.

Hodja had counseled that when we reached Paris, we were to make our way from the Gare du Nord station to a park, where, just as in Rome, refugees gathered. We hoped we could make contact with people, possibly even agents, who would help us get to Calais.

I had half a mind to ask a passerby the way. I hoped I could manage to communicate using a mix of sign language and English. But the Parisians who brushed past shot us looks as though we were something unpleasant to be avoided—a nasty smell that polluted the Parisian air, scented or not.

An older man, with the instantly recognizable craggy features of a life spent in the Afghan weather, squatted near one of the station entrances, rattling a paper coffee cup at the rushing commuters.

"*Salaam*, *Kaka,*" I said in Pashtu. "Where is the park, please?"

"That way," he said, pointing to a busy intersection. "Seven blocks away. But don't expect too much help, boys. It's every man for himself."

He dismissed me with a shake of his cup, wrapping a blanket around his shoulders more tightly about him. It was cold, and I was beginning to shiver now, too—I was grateful for the jacket Hodja had bought me. The gray sky was heavy with fat clouds.

"Come on," Shafique said. "It's going to snow."

I feared he was right as I stamped my feet to shake some blood into my cold toes. I felt wretched—exhausted and hungry. After twelve months of traveling, this was the reality of my existence, and the persistent discomfort didn't make it any easier. Lingering in my recent memory was the fact of the days spent in the children's home, where I had been clean and fed and had slept in a warm bed. The memory seemed to mock me.

Tiny little flakes of something between rain and snow started to fall from the sky as we shuffled in the direction we'd been given. I should have felt excited: Paris meant Great Britain was tantalizingly close. Just a little further, and I would be there. What "there" meant in reality, I still wasn't sure, but all that mattered was that Hazrat was there, and I would be running no longer.

"Gulwali, look."

I followed Shafique's gaze.

Three other Afghan-looking men had rounded the corner just ahead.

"Brothers," Shafique yelled over the road. "Where is the park?"

"Around the corner and you are there," one shouted over the traffic. "But you had better hurry—the charity people only serve the food until six."

Food. We broke into a run. We hadn't really eaten on the train, and we'd only had snacks in the Turkish cafe.

We just made it in time, as they were packing up. A few glorious minutes later, I was bathing my face in fragrant steam. The smell of white beans and chicken made me drool.

Shafique had almost finished his stew already.

"It's too hot to eat. You must have a stomach of iron."

"It's too good not to," replied Shafique, wiping broth from his lips. "*Très bien, monsieur.*"

"Nice French."

Small groups of refugees were scattered around the park. A short line spilled from the tin shed that served as a soup kitchen.

"You finish your food," said Shafique, throwing his empty plastic bowl into a bin. "I'm going to see where we sleep."

We had heard about a homeless shelter where they would let refugees stay overnight. Hodja had said that because we were so young, we might get lucky.

I was mopping the last juices up with a piece of bread when he returned.

"Bad news."

"What?"

"The shelter is closed."

This was a blow.

"So what do we do?"

"I'm not sure," said Shafique. "Some of the others are sleeping together at the far end of the park. We can go and join them."

"In the snow?" I was inwardly groaning.

"It's not that cold," Shafique said. "Look, it's barely even snowing now." I had begun to love Shafique for his optimism, which I knew was for my benefit. "There's a tree over there," he went on. "It's practically dry underneath, and the branches will keep the weather out."

We tried to get comfortable beneath our shelter. It smelled of ash and damp. Shafique was right: the earth beneath the tree was almost dry, but that was little cause for celebration as the temperature dropped over the coming hours, bringing bigger, fatter flakes of snow with it. We lay back to back, trying to preserve our body heat.

Shafique was soon snoring. My eyes got heavy a handful of times, but there was no way I could sleep. The cold and the strange sounds of the city kept me awake—sirens and traffic noise, voices calling out through the night, as though no one was sleeping and the great mass of humanity of the city was continuing to go about its business as I and the other homeless refugees lived like a subspecies.

I had had such high expectations of Paris, the city where perfume rained from the skies. And yet all I had witnessed was a dirty, smelly, and cold city, filled with Parisians who shied away from us in horror.

I got up to go to the toilet. I guessed it was around midnight. My hips and shoulders ached with cold, and I limped with pain as I tried to get my blood moving.

I looked for somewhere to relieve myself. It felt good to be moving after the cold ground, so I wandered in no particular direction. Piles of bodies lay dotted around the park—mostly on benches, sheathed in cardboard. The low glow of a telephone box hung in the distance, snowflakes caught in its light like moths. It almost looked warm. I let its glass door slam shut behind me, and my breath began to fill the space.

I leaned against the steel case of the telephone and closed my eyes.

Fleetingly, I imagined using this phone to ring someone—perhaps my mother. But I realized how stupid that was. She didn't have a phone, and I hadn't been able to contact her once on this journey.

A group of drunk young Frenchmen rolled past, talking boisterously and waving their glowing cigarettes around with extravagant flourishes.

I shrunk down, trying to hide. They passed without seeing me.

I dozed for a little. I was warming up a bit now, as I crouched in the bottom of the phone box; despite being forced to crouch with my arms across my chest, this was the most comfortable I had been all night. A snow-covered lump rattled along the street toward me, rousing me. A man in a heavy coat and hat pulled low shoved along a shopping trolley, his little dog trotting beside him. I watched him, puzzled. He didn't look like a refugee. I had seen other such homeless people through my journey, especially in Rome and Athens. I couldn't understand how this happened in European capitals.

The aluminum edge of the door suddenly crashed into my right ankle. Two black-clad figures loomed over me, the muzzles of their guns staring me down. For a second I thought I was back in Turkey or Iran.

Police.

I said nothing, frozen into silence.

The shorter one summoned me with a black-leather-clad finger, indicating that I should come out.

I rubbed my shoulders to try to make them understand I was cold and had nowhere else to go.

"*Halas,* finish," the taller shouted, and shoved me down the street. "*Yallah,* come."

These were basic Arabic words. They were obviously words he thought all refugees might understand.

He shooed me away like a dog, making a *tksss tksss* sound with his tongue and teeth.

I put my wrists together and held them up. I was trying to gesture to them to arrest me. Right then, a warm cell seemed preferable to dying of cold.

"Go. Move."

I skulked away like the animal I felt, retracing my fading footsteps in the snow. I made my way back to Shafique. He was still sleeping, but now shivering uncontrollably. I lay down next to him and tried to sleep.

When I woke up, Shafique was still out. I sat there, trying not to shake. I began to move around for warmth, eventually crossing the road to a sort of fire hydrant with a tap on top. People were washing their face and hands in it. I did the same. The temperature of the water sucked the air out of my lungs.

About an hour later, the soup kitchen opened for breakfast. Shafique and I staggered over to join the queue of dozens of men, all shivering and moving about to try to warm themselves. Shafique was still shivering, his teeth chattering. I don't know how he'd managed to sleep.

At the soup kitchen shed, a sandy-haired Frenchman with a neat beard served me hot tea and a baguette with cheese. Gratefully, I cupped my hands around the mug of black tea. It was bitter and hot. I could feel it working its way down to my stomach, warming me from within. I loaded my cup with sugar until the lady behind the counter cleared her throat. Shafique savaged his piece of long bread.

"This is good," he said, cheeks bulging.

"How can you be so happy? We nearly froze to death."

"Yes, but we didn't, did we?"

I took a bite of my bread. "Let's find out what to do. I don't think I can cope with another night here."

We went over to where a group of refugees was huddled on boxes. The cardboard sheets were laid out in the same way that kilims and cushions would have been back home. This was the sunniest part of the park, and everyone was trying to get a little warmer from the few rays of wintry morning sun.

As we sat there, people who had been sleeping in different parts of the park, or the lucky ones from the shelter, came to join us. It took on a slightly jovial atmosphere despite the misery, reminding me of home, where people would congregate in the bazaar after breakfast, sitting with their backs against the wall of the shops as they caught up on the local gossip and politics of the day.

Shafique was talking to a fellow Afghan in a black woolen coat and jeans. I shuffled over to join them.

"Gulwali, this is my friend, Jan. We traveled together from Greece to Italy. I lost him in Rome, and here he is."

"Pleased to meet you, Gulwali." I shook his warm hand. "Shafique told me about your night."

"Yeah. It was what it was," I said, easing myself to the ground.

"Have some blanket," Jan said, pulling a grubby white polyester cloth from beneath himself.

"Thank you."

Jan looked at me. "I feel like I know you from somewhere, Gulwali. Like I've seen your face at a wedding, or something. Where are you from? Who was your grandfather?"

I wasn't thrown by this. Afghan men asking about lineage was a perfectly normal thing—in Afghanistan, we did it all the time. It was how we identified people.

We traded relatives back and forth until we worked it out: my grandfather and his father were cousins.

I laughed: that meant we were cousins. I was picking up a few of those on this journey. I thought briefly of Jawad, and wondered if he'd managed to leave Istanbul yet.

"Nice to meet you. So what brings you here to Europe, cousin? Business or pleasure?" I joked.

"Ha-ha." His laughter was genuine. "But I should be asking what *you* are doing here. You're so young."

At that, my humor evaporated. I stared at my toes as I tried to find the words. "My mother sent me and my brother, Hazrat. It was too dangerous at home. My father and so many relatives are dead—killed by the Americans."

"I heard. I'm sorry—I got the news when I was in Peshawar. May Allah rest their souls in peace."

"Thank you."

"So where is your brother? Is he with you?"

"I don't know. I think he is in the UK now. I want to get there to be with him." It hurt me to say those words.

The events of the past drifted through my head like a strange dream. I think the others felt the same way, because we just sat in silence for the next few minutes.

"Are you boys hungry?" Jan clapped his hands together. "I know a nice place that is not too expensive."

The three of us walked to a Turkish cafe, where Jan bought us dinner. It felt so normal that for once we weren't looking over our shoulders, worrying if the police might appear. The locals went about their business without noticing us, and the man who ran the cafe served us kindly.

"So, what is your plan now?"

Shafique's cheeks bulged with chicken, so I answered for him: "Shafique hasn't decided what he's going to do. Maybe he can stay in France—?"

"What about you, Gulwali?"

I knew what I was doing. While my family had only paid for me to get as far as Greece, I had learned enough from Hodja to realize that if I could make it to Calais on my own, I could figure it out from there.

"I will take a train to Calais."

Jan raised his eyebrows. "By yourself? You've heard how dangerous Calais is, right?"

I knew how it was—Hodja hadn't held back. And even that morning, when chatting to people in the park, some of them had recently returned from there because conditions were so bad. But I didn't see that I had a choice. And I didn't want Jan to think I was a little kid.

"Yes. So what?"

He raised an eyebrow at me. This made me annoyed. "I've heard the stories. But so what? I'll go by myself if I have to. I can find help when I get there."

The night before had made me more determined than ever to push on quickly. England felt close now. I was lucky that I wasn't in a Parisian jail, that the police from last night had simply moved me on.

"Why don't we go together, cousin?" Jan suggested. "It was my plan to leave in the next few days anyway, but as you are here, I can come with you. I'm not happy about you going alone."

His protective nature was getting on my nerves, but in a way I was touched. He saw me as family, and it was his duty to be there for me.

We tried to persuade Shafique to come with us too, but he insisted he wanted to wait in Paris a while longer. He had heard of a good agent that could arrange everything from there. He had had bad experiences of doing things without agents further back in his journey, and didn't want to risk it again.

"But don't you trust me, Shafique? I can sort it out for you. I got you to Paris, didn't I?"

"So why don't you stay here in Paris with me, and maybe in a few days we will have found an agent and we can move? Tonight we can find this shelter and have a warm bed," he countered.

I did consider it, but I didn't want to. As much as I liked Shafique, I thought he was a bit naive. He had only been on the road for six months, half as long as me. He'd got to Paris far faster than I had, but it meant he hadn't learned as much. That morning, as we had chatted to the others, he had been trying to find agents. I was worried he was going to get ripped off or tricked.

I had about 350 euros left: I'd had to buy tickets and food for Shafique and me. Hodja had given me 150, and the agent had given me 300 in Greece—the supposed change from what my family had paid up front. I also had about $100 left. But Hodja had warned me Calais could be very expensive and I would need every penny. I didn't want to risk losing it to some dodgy agent here. Far better, I thought, to get there first and find out the lie of the land.

Shafique and I argued for a while longer. He kept trying to get me to stay, and Jan and I kept trying to persuade him to come with us to Calais.

It was no good.

Jan and I went to the train station to get two tickets. The next train was early afternoon. The train to Calais went from a different station, but I had no idea where it was. We walked through Paris trying to find our way.

Surprisingly, it wasn't that hard—there were Afghans all over the city who helped us on our way.

CHAPTER 26

BY THE TIME WE FOUND THE STATION, THE TRAIN WAS ready to pull out. Jan and I sprinted along a polished platform. I realized now that I was relieved that he was coming too—despite my bluster, deep down I hadn't wanted to go alone. We boarded and snoozed comfortably during the ninety-minute journey.

We arrived just after lunch. We weren't the only foreign faces in Calais: dozens of us forced our way through the melee of the station. There were refugees everywhere. Filthy bodies were sitting wherever they could; others stood around in little groups.

Jan approached a huddle of men. They were so dirty we couldn't really tell what nationality they were.

"Excuse me, friend . . ." he started.

"You newcomers need to go to the Jungle," said a skinny, gray-haired man. He spoke to us in Pashtu, but his accent sounded a little odd.

"You from Afghanistan, brother?" asked Jan.

"Pakistan. Waziristan," he said. I could hear it in his accent now he'd said it.

"Is the Jungle so bad?" I asked hopefully.

I had heard stories in Rome of how awful the Jungle was—the name given to the port in Calais, the place where refugees gathered and lived in order to try to get on board trucks to England. But I also suspected it might not be as bad as people made out; I thought maybe the details were exaggerated to discourage people from going here.

The man looked at us. "Whatever you've heard about the Jungle, it's worse than they say. The only thing you need to know is this—they don't want us there. The West loves dogs, almost as much as it loves war. Bush and Blair consummated their invasion, and we are the unwanted puppies of their bombing. They don't want to let us in to the warmth of their fire—but they don't have the stomach to kill us. So, here we are, locked out in the rain and cold, fighting over whatever scraps fall from their table."

"What do you mean?" I asked.

"The people there are scum. They are thieves and liars. And the French police treat us like animals. To be hunted and chased around and around, like it's just a big game." He let out a bitter little laugh. "This is why they call it the Jungle, boy."

"So, why should we go, then?" I wanted answers.

"Because, my boy, you have nowhere else to go. It's the same reason all these miserable creatures are here. They've come this far and now they cannot stop."

"How long have you been here?" Jan asked.

"Six weeks."

"So why did you not get onto a truck?"

"Oh, I've got on lots of trucks. Every day I am catching a truck. But then the driver, or the police, or the security people, find me and remove me."

"They don't arrest you?"

"Like I said, to them it's just a game. If they arrest me, then they have to put me in a nice warm cell, give me food. No, they prefer to dump me on the side of the road and make me walk back to this miserable place. I think they want to teach me a lesson."

Nothing this man was saying made sense to me.

"A lesson about what?"

"I don't know. Maybe they want to teach me that I was a fool to flee my home. That I would have been smarter to stay among the fighting. That it is better to watch your family and neighbors die than walk mile after mile along freezing roads with an empty stomach."

"I never did like school," said Jan, trying to lighten the mood.

"Then you will hate it here," the man said.

Jan and I stared at each other, and then back at the man.

"So," said Jan. "Where is the Jungle? I can't work out which way it is."

"Over there." The man pointed down the street. "Follow that road for an hour. You will smell it before you can see it." He sighed. "Do you have an agent here?"

We shook our heads.

"No one does when they arrive. I wouldn't worry. Money talks. Take my advice and ask for the Kurd they call 'Le Grande Fromage.'"

"The what?"

Jan burst out laughing. "It means 'Big Cheese,' Gulwali."

Laughter erupted from my throat too. Big Cheese—it was the most ridiculous name I'd ever heard.

"Kurds," said Jan. "Forget Bush. I am beginning to think it's the Kurds who rule the world."

WE'D JUST STARTED WALKING IN THE DIRECTION HE'D TOLD US when we spotted a couple of women serving some immigrants tea from a large urn. We went and got a cup. A woman in a wool beanie handed me a steaming cup with a smile. It was nice to see a friendly face, even for a moment.

I saw familiar faces from the various countries and places I had stayed along the way—no one that I knew, but faces I recognized. It never ceased to amaze me that we could have crossed half the world, yet Afghans always seem to bump into each other. I supposed it was not too surprising given that we were all ultimately heading for the United Kingdom.

I knew that for the majority of the people there, the UK was their destination. It's the idea, the notion that, because it's the hardest to get to, it's the best place, the last stop, the end of the road.

The end of the game.

That's not the only reason, of course. Afghans have historical and cultural connections to Britain, either through the former empire or through war. Or language. Many have family already there. Also, the UK was seen as tolerant and fair to immigrants, unlike some European countries such as Bulgaria, Hungary, and Turkey. Still, I was anxious about loitering around the station. I felt vulnerable, exposed.

"Let's find some food," I said. No matter what lay ahead, a full stomach would help.

Other refugees told us there was an evening soup kitchen twenty minutes' walk away. They were going that way. Apparently, the French police didn't make arrests near the various food distribution points, which were run by French charities, so these were safe places. The refugees tended to hang about at food places for that reason. They told us they also walked around in large groups. If the police spotted you alone or in a pair, they were more inclined to detain you; in a larger group, you tended to get left alone.

The line for dinner stretched for hundreds of meters. Jan and I stood in silence as we queued for about half an hour, and I watched the tide of people around me. The few women there stayed close to their male companions. I saw the way some of the other men looked at them, and it made me angry to see this. These poor women looked scared enough, without strange men staring at them in a bad way. A few of the women had small children with them—filthy, snotty-nosed little urchins with rattling coughs. My heart broke to see them. They kept their gazes low, seizing what little dignity there was to be had.

At the front of the queue now, I could see four older French women stirring the huge pots needed to feed so many hungry mouths. I think perhaps it was run by a local church. The women dished out ladle after ladle of thick, white paste. It was rice, but it looked nothing like it, while the lentils that went on top tasted of nothing. All the same, I ate it gratefully. Around me, about two hundred others were doing the same.

Some Afghan refugees made their way through the crowds, offering to guide people to England—for a fee, of course. They were salesmen working on commission for the agents, and they were very good at spotting the new Afghan arrivals. One by one they approached me and Jan.

"You are new."

"I can help, I know a good agent."

"I know a good Afghan agent, our own people are more trustworthy than the Kurds."

We were suspicious of all of them, and determined to find an agent called Karwan. Hodja had told me about him and said he was the best of the bunch because he had a high success rate. He had also advised me to bargain hard.

We asked around and found some refugees who were "his" people. They told us they would take us to him.

We walked for an hour or so with these people, slowly making our way toward the port area. Nothing could have prepared me for what we saw as we walked. It looked like the world's toilet had been flushed and the mess washed up here. All along the way, migrants and refugees cluttered the roads. Some were in small groups, walking with purpose to an unknown destination; others merely sat or lay where they could find a comfortable space. Face upon face was deeply etched with hardship and misery.

"There's no escape boy," a voice said to me in Pashtu from beneath a sleeping bag. It looked like it had been red once, but was now a sickly shade of green-brown. I didn't have a reply, but it did leave me feeling that as close as I was to Britain, I might be more distant than I had ever realized. All I could smell was human decay and the diesel fumes from the stream of double-trailer trucks that crawled past us.

On the other side of the road were rows of warehouses made of corrugated iron, with barren swathes of concrete between, all surrounded by huge fences. Impenetrable spinneys of coarse seaside scrub and willow trees made the place disorientating and bewildering. Black plastic shopping bags whispered from every sharp twig and thorn.

The place was so vast and so confusing, it really was no wonder they called it the Jungle.

CHAPTER 27

As we neared the jungle, I began to notice sapphire-blue tarpaulins everywhere. Each one marked a makeshift house or communal shelter, which were interlinked by a maze of tracks in the sandy soil. It was a temporary city of desperate human flotsam. There were huge piles of litter and human waste everywhere, dumped in black plastic bags that were stacked as high as a man. White plastic water drums, fertilizer sacks, plastic cups, food wrappers, bottles, discarded clothing, and worn-out shoes—an abandoned, accumulated mess.

We passed a small home built from a patchwork of three damaged tents—black, green, and orange. A chimney made from discarded air-conditioning ducts peeked from the roof. Three Sudanese men in long black jackets with their hoods pulled up sat around a small fire, trying to boil eggs in a broken saucepan. A moldy beige leather armchair, a broken office chair, and an upturned supermarket shopping cart functioned as furniture.

Jan and I nodded a greeting to them and continued walking through what was essentially their living room, taking care to step over a shallow channel that had been cut across the ground to drain away rainwater and raw sewage.

Our guides took us to a group of refugees huddling around a fire. An old man threw a plastic bottle into the flames, which belched out clouds of sickly sweet chemicals as the plastic melted. At least it was momentarily warm.

A muscular man appeared.

"Gulwali, this is Karwan."

We shook hands. His Farsi was good enough for me to understand him, but his Kurdish accent was still strong. He wore a shiny black ski jacket and faded jeans. Wraparound sunglasses sat on his head, holding a long sweep of black fringe in place.

"You are the newcomers. I am told you want to get to England?"

"Yes," I said. "Tomorrow."

He scratched at the patchy black stubble on his chin. "Well," he replied, "have you put any money with anyone?"

I knew what he meant. He wanted to know if I had any money lodged with one of the shadowy "third party" individuals—in other words, did I have credit to travel? Most refugees reaching the Jungle have not paid in advance, as the national agents in places like Kabul and Pakistan only promise to get you as far as Greece, or maybe Paris, and that's all they take money for. After that, you need to use one of the specialist Calais agents, people like Karwan, to get you across on the ferry. This work—getting people onto trains and trucks—was all they did. The cashless people in the Jungle tended to pay for this part of the journey by having a third party somewhere in France or the UK act as guarantor. The third party was usually someone known to the agent and, more critically in possession of a European bank account. If you already had family or friends in Europe, they could make the contacts on your behalf. There had even been people in the park in Paris who had promised they knew third parties who could help with this—if you paid for it in Paris (or at least deposited the funds), then in theory was that by the time you reached Calais the specialist agent would be expecting you.

Shafique had remained behind in Paris trying to do just that. He had wanted to pay up front and sort it out in advance, not turn up here without plans like Jan and I had done.

The other option on offer in Calais was one where the agents would give you details of his own contact with a bank account. These people would be complete strangers to you but they were still willing to guarantee for you, once they took some background details—the names of your family, and where they lived. The idea was that once you made it to England, you paid them back in full. Of course, you could have just disappeared and not paid them, but in reality this rarely happened. The strong tribal networks we had back home stretched all the way into Europe, and if you didn't pay it was a safe bet that the agent and their guarantor would hunt you or your family down.

It sounds like a crazy system but it works. If it didn't, it wouldn't happen.

Karwan looked at me expectantly. "I have contacts in the UK who can arrange your credit. I can give you their number and a phone to call them. You make the agreement and once that is arranged, I will ensure you reach the UK. Guaranteed. Five hundred euros each."

He scratched his chin.

I stayed quiet and scratched my chin too. *Guaranteed.*

"We have cash."

"If you want to pay me cash, pay me three hundred euros each."

"That is too much. Where are we going to get that sort of money?"

He didn't answer.

"We don't have that much money."

He continued to scratch his chin, this time with an air of annoyance.

"I can give you fifty euros," I said, taking a thumbed note from my pocket and waving it at him. I had another fifty in there too, but I was determined to not let him see it.

He stared coldly. "Don't waste my time with fifty euros."

"Please, Karwan. You've got to help us."

My hands wrapped around his wrist and I looked up at him, willing my eyes to fill with tears.

Under normal circumstances, Karwan might have pushed me away, or even hit me. I think he wanted to. I could sense the tension in his arms, but it was a risk I was willing to take.

"*Please,* Karwan." I looked up into his eyes again.

"So, what about the fifty euros you still have hidden in your pocket?"

He'd seen. Reluctantly, I gave him the money.

He looked over at Jan. "And you. What have you got?"

Jan handed over another fifty.

"Please, Karwan. Take us."

He shrugged his broad shoulders. "All right, I'll take you," he huffed, fixing me with a glare that told me he knew what I had done. "But if anybody asks," he hissed, drawing close to me, "you'll tell them you paid five hundred pounds each. Into a British bank account. Got it? If you tell anyone the truth, I will kill you both."

I WAS FACEDOWN IN THE FREEZING MUD, SCRAMBLING BE-neath a gap in a wire-mesh security fence.

Karwan leaned his weight back to hold the mesh up. "Hurry up," he urged, the strain showing on his face.

Jan, I, and four other Afghan men wiped ourselves off as well as we could and stared around at the large, brightly lit parking area filled with trucks. Through the sleet I could make out dozens of figures swarming around the backs of the high-sided trailers, or bowing beneath the wheels as if in search of an unknown goal.

"Come on," said Karwan. "You don't want to miss the best spots, do you?"

We jogged behind him. Groups of refugees, mostly in ethnic packs—Asian, African, Arab—roamed the park searching for any opportunity to board.

"Check underneath," Karwan said. "They are good spots, especially for a little guy like you, Gulwali."

I tried to check under one trailer, but a hand caught my shoulder. It belonged to a Pakistani man. His dark eyes peered out from between the folds of his balaclava. "Don't listen to him, boy," he said in Pashtu.

"It's too cold to ride underneath. If you don't freeze to death, you'll lose your grip and get crushed. Better to ride in the back."

As our search continued, it became clear that there were simply too many people for the few hiding places.

"Let's go," said Karwan. "I have a better plan."

Two hours later, Jan and I sat shivering back to back beneath a concrete motorway bridge. It was long past midnight.

Karwan had gone on to make "other arrangements." He returned soon after, brandishing heavy gray bolt cutters. "The keys to England, gentlemen," he said, waving them at us.

He cut our way into the parking area of a ferry terminal.

Unlike the first parking area, this one was busy with truck drivers. The three of us lay on the ground, trying to get a sense of the activity. Many drivers checked the doors on their trailers and searched the chassis. Others chatted, drank coffee, and smoked cigarettes. Cars full of families and businessmen formed a series of orderly queues nearby.

Night was giving way to day now, and I could feel desperation rising inside me. Karwan yelled to us to run with him across the open car park, looking out all the while for the police or the sudden appearance of a driver.

We crouched beneath the tail of a truck.

"There." Karwan suddenly pointed toward a silver estate car parked to one side. The driver's door was open and we could make out a figure standing against the fence, steam billowing from around his feet as he peed.

We stalked forward, mindful of the heavy slopping noise of our footsteps. Jan quietly opened the back door. The passenger seats were full of suitcases and boxes, but the footwells were empty. My heart was racing. I slid into the tight space, crawling into the hole behind the driver's seat in an effort to make room for Jan. He squeezed in behind me and I heard the door click shut. I barely dared breathe as the driver returned from relieving himself.

My legs were beginning to cramp in such a small space, but the car was warm and I was thankful for that. Before long, we started to move forward. Jan patted my ankle as if to congratulate us—we were on our way to England.

The driver crept forward a few feet at a time, occasionally lighting a cigarette or adjusting his car radio. Suddenly I tensed, aware of voices at the driver's window. Flashlights probed the car's interior. A shout.

The driver turned, following the beam of light. "Hey," he bellowed, throwing himself out of the car.

The door beside my head swung open and a strong hand dragged me onto the cold ground. Police stood over me, shouting.

Jan was on the other side of the car. "Leave him alone, he's just a boy."

The police shouted back in angry French. The driver, who had now recovered some of his composure following our discovery, shouted in my face and pushed me against his car before being held back by the police.

The police marched Jan and me down the line of cars, back toward the main entrance, near the ticket barriers.

"*Halas,*" they said, waving us away into the night. "*Allez.*"

We skulked away into the darkness.

"Why didn't they arrest us, Jan?"

"I don't know. A cell and a hot meal would be welcome right now, wouldn't it? Maybe that's why not."

We found the train line and began the long, cold walk back to the Jungle. It was snowing more heavily now, and the freezing wind stuck my wet jeans to my legs, sucking the heat from my body. My stomach—knotted with tension for so long now—relaxed a little, so that the pain of its emptiness cut into me.

By the time we got back to the Jungle, I was exhausted. We lay down on some sheets of cardboard and tried to sleep before Karwan came for us in the early evening to attempt it all over again.

And that became our routine for the next month.

WE SOON LEARNED THAT EACH SMUGGLER HAD THEIR OWN area—a section of a car park or rail line that was exclusively theirs. Sometimes we'd see several groups of refugees all trying to board vehicles, yet no one wandered into another smuggler's area. It was, despite the chaos, very organized.

Occasionally the smugglers would settle their differences in a brawl. But most of the time we were running—running across car parks to stationary vehicles or running away from the police or drivers. It was easy enough to get on, in, or under a truck or vehicle. It was impossible to go undetected by the police and border guards.

They had dogs and cameras—and they knew all the best places to hide, just like we did. Whenever we got caught, the police would just let us loose again—providing we were a long way from the Jungle. If not, they took a vicious pleasure in driving us to a remote location and dumping us at the side of the road. They didn't want to take our liberty; they were only after our time and precious energy so that we stopped trying. In reality, this only made us try harder.

Under Karwan's careful guidance, our little flock developed an economic hierarchy. Jan and I were always last to board when it came to enacting Karwan's nightly schemes. We had paid the least, and so we were his lowest priority. Eventually we blackmailed him.

"Stop putting us last, Karwan," I complained. "We can see what you're doing. Stop it, or I'll tell the others how much we paid."

Karwan seethed at me, but Jan stood his ground.

"Do as Gulwali says, Karwan."

He did.

"Karwan's Guys," as we now thought of ourselves, formed our own small camp, and we started to settle in as best we could. Jan and I befriended a quiet and thoughtful Afghan. I nicknamed him Qamandin—

"leader," or "man in charge." He was only a teenager, but he had a strength of character that we instantly warmed to. We made a little house using a tarpaulin and the branches of a willow tree; Jan and Qamandin found some pallets on a construction site after a long walk home one morning. The tarpaulin had been claimed in a similar manner. So too had a gas bottle, liberated from an unlocked shed in a beachside neighborhood. The theft didn't sit well with me. But we were acting out of hopelessness. It was a simple choice of steal or die.

Life became a soul-destroying cycle of escape, capture, theft, and constructing our shelter. Only the physical necessities of eating at charity food points, the occasional wash, and broken sleep disrupted the cycle.

The Jungle was ethnically divided into ad hoc nationality areas. The Africans seemed to be the poorest and slept nearest the car park fence. They didn't seem to have agents and appeared to be trying to get on trucks themselves. There were also Chinese, who only seemed to turn up at night with cans of beer to sit and talk to us. They didn't seem like refugees. They may have been migrant workers living locally—there were lots of factories nearby the port.

Regardless of nationality or intention, everyone tried to make the best of the situation. Someone found a tattered football in a playground, which created a few precious moments of fun.

During the day, the police loved to raid the Jungle, probably because they knew we were likely to be resting. It was common to have my charity-donated blanket pulled from me, a screaming police officer shouting in French in my face. They would move us on, beating any guys who resisted. We would then be forced to stand shivering in the cold while they questioned us about things we knew they didn't want to hear answers to. Of course, every time I made sure to tell them I was only 13, but they didn't care.

Sometimes activists came to the Jungle to form human shields between the police and ourselves. They tended to be youngish English and French men and women, although there were a few older people involved as well. They would stand outside our shelters shouting at the police, demanding they stop harassing us. They had loudspeakers, which they used to try to embarrass the police into leaving us alone.

We were grateful for their empathy and support. Some of the women used to like to hug me. If felt nice to have a moment of human comfort. Sometimes they brought fruit or other food with them—that was even better. But we knew they weren't going to be here all the time. And so did the police.

Sooner or later, the police knew they would get their way.

A couple of times, officers detained me for twenty-four hours. I didn't mind—it was warm, and usually it offered the chance to see a doctor. There was a volunteer clinic where a young French-Afghan doctor helped me out. My face and body were covered in pimples—a consequence of poor hygiene—but there was little they could do. The medical clinic had showers, although you were only allowed to use them if you had a medical reason to do so, but the doctor was nice and always let me sneak in for a wash.

I lived in the same clothes for weeks at a time. I wore them until they were filthy rags—or until they became so infested with fleas and lice I couldn't stand it anymore. Fortunately, the charities that ran the food places were often giving away secondhand clothes. That became my definition of a good day—a hot meal, some new clothes, a visit to the doctor, and an illicit shower from which I emerged clean and dressed. I hated the filth. It gave me some small pleasure to throw my old rags on the fire, and watch as the bugs popped in the flames.

The humiliation was hard to bear. Many of the faces I saw spoke of the same thing. In their own countries, many of these people had power, even the respect of their communities. Here in the Jungle we were barely human. We were the beasts that gave this place its name.

I imagined myself running up to some high-ranking French official and shaking them to demand answers. It wasn't my fault I wasn't born in Europe. My home was a war zone—did that somehow make me less human?

I spent a lot of time just wandering the Jungle or the food area near the train station, which helped pass the day. The police would see me and often assume I was out stealing—pickpocketing or shoplifting, and, admittedly, I sometimes was. They would drive past slowly, making a point of staring at me. Every so often an officer might take pity on me—I suppose some of them had children or relatives my own age. They would give me a few euros or perhaps some food. I always accepted it with a smile.

There was a routine to my life in the Jungle, though I was loathe to accept it. Every night, regardless of how futile our attempts were, we all tried to get to England.

On one particular night, two weeks after arriving at the Jungle, Karwan selected Jan, Qamandin, me, and twenty-one others for a high-sided truck. The driver had left the cab to use the toilet.

"I've already cut a hole in the top," Karwan told us proudly, as we lay in the grass on our stomachs, shivering on the cold ground.

When we got to our feet finally, I was pleased to be moving. But when we approached the truck I wasn't so sure. "What are those?" I asked, pointing at red signs with skulls on them on the truck's side.

"Nothing to worry about, Gulwali," said Karwan. "Just don't eat anything you find inside. It's *haram*." He boosted me up. "You first, little one."

I fell headfirst through the hole and into the darkness, where I landed heavily on what I thought was grain or beans. A cloud of soft powder blew into my face. I was sure it was agricultural. I crawled into a back corner and stayed still. Whatever this stuff was, it was soft to lie on, which made a nice change.

I heard the others climb on board and find their positions. The twenty-four of us stayed in the truck all night, not making a sound. In the morning the driver came and the truck started moving. We felt it go in the direction of the port but, just before it should have turned right toward the port, it drove straight on for half a mile instead. You get used to judging distances, so we knew something wasn't right.

When the driver opened the door we ran, because we knew it wasn't the port.

He was screaming at us in English, only a few words of which I understood: ". . . not run . . . ambulance, ambulance . . ." None of us had any idea what he was carrying on about, so we just kept going as fast as we could.

I was black from whatever had been in the truck, covered in fine dust. Maybe because I was the first in the truck, but for whatever reason, I had emerged far dirtier than anyone else.

We found a water pipe to wash some of the mess off, and I rubbed water over my face. It felt good to get it off. I felt clean again. I felt . . . burning. The pain intensified. It felt like my face was on fire. I screwed my eyes closed. "Help. Help. Help." I shouted, scrambling for more water.

Qamandin was there, splashing me down. "That's it, Gulwali, wash it off."

"Get my eyes. Rinse my eyes."

More burning. The water only made it worse.

I screamed—the pain was beyond anything I'd ever felt, as harsh chemicals corroded my skin.

Qamandin dragged me like a dead goat, all the way past the food points, toward where the medical clinic was.

On the way we passed the food point. Thank God the nice lady in the beanie hat was there. She saw me and screamed in horror. "We've got to take him to the hospital. Hurry." She grabbed hold of me and put me in her car right before I passed out from the pain.

WHEN I AWOKE, EVERYTHING WAS BLACK. ALTHOUGH I COULDN'T see it, I was in a real bed. The sheets were soft to the touch, and smelled clean.

"How are you feeling?"

It was a woman's voice. I thought I might be dead, but she had a French accent, therefore unless heaven was in France I was most likely alive. A hand rested on my shoulder, then lifted, and began to peel the bandages from my head. Slowly I could detect light. I felt gauze pads being lifted from my eyes. A blurry image hovered above.

"Gulwali," said the blur.

A bright light blinded me—first in my left eye, then the right. I blinked and the image grew clearer. A middle-aged woman stood over me—blond hair, white gown, kind face.

Next to her was an Afghan man I didn't recognize. He stepped forward and addressed me in Dari. "You are in Paris. In a hospital."

I just blinked.

"The doctor says your eyes will be fine. They will give you drops for them, and cream for your skin, as well as some painkillers."

I smiled. My lips hurt. They felt blistered.

"We will discharge you now. We will take you back to where you want to go."

Where I wanted to go? I had nowhere, only my temporary home in the Jungle.

The journey back was very strange. I lay in the rear of the ambulance, my face misted with ointment. The Afghan rode with me, but we said nothing. My mind was filled with relief about my eyesight, but I was equally concerned about what this new setback might mean for the rest of my journey. I was also furious at Karwan for making me get in the truck. He must have known, suspected at least, that it was dangerous. Was that why he let me go first?

The rights and wrongs of the matter were no longer important. The only thing that mattered was my getting to England—by fair means or foul.

WHEN I RETURNED TO THE JUNGLE, THINGS HAD CHANGED. Karwan had disappeared. The agents were so secretive. The police were always asking us to reveal who they were. The truth is, we didn't really know. Secrecy was part of their business. They didn't reveal where they lived, or any other details about themselves. I couldn't figure it out—but it wasn't of major concern to me, anyway. I was paying them to get me to England, not be my friend.

We found a new agent, a man called Pustiwan. We begged him to take us on. We didn't have enough money, so we had to call one of his people in the UK to guarantee credit.

I found myself getting tired very quickly after I got burned. I still don't know what chemical had burned me, only that it had some kind of chemical reaction with water, triggered when I splashed my face.

Jan and Qamandin were kind and looked after me, bringing me food and insisting I stay in bed and rest. But when we went out to get on the trucks, the other people under Pustiwan's charge got very angry with me, accusing me of being too slow and spoiling their own chances. They looked at me now as if I were a weak link—as if what had happened on the chemical truck that night was somehow my fault.

It just added to my sense of despair. I'd been there for so long, and I barely knew a single person who had made it to England. I prayed that we'd get news that someone had made it, just to give us hope. I was beginning to wonder if this was what the rest of my life would look like.

When I felt stronger, we tried jumping on trains heading for the tunnel. But it was just too dangerous—I knew the risks of trains first-hand, and even if you got on board, it was easy for security to find you.

And that meant another long walk back to the pile of stinking rags we called home.

Over the time we were there, security became tighter, too. Some guys were detained and kept in jail for up to twenty days.

I'd got word that Shafique was in Dunkirk—he thought it might be easier to cross from there. I'd thought about going myself, but as bad as things were here, I couldn't quite believe there was an easier option that I'd been overlooking. I was miserable, stressed beyond belief. I was a failure for coming all this way, only to be thwarted by a stretch of water I could almost picture swimming on a calm day.

One evening, the others went out. I don't really remember what they said they were going to try. I just felt sick and tired. My face was still incredibly painful; it had blistered, turned black, and was peeling. I wished Jan and Qamandin good luck, and buried my head under my dirty blanket.

They didn't come back.

I wanted to be happy for them, but it was the worst thing I could imagine. My one chance to escape had come and gone because I had been too tired. I cursed myself—told myself what a stupid little boy I had been. When I thought I couldn't feel any worse, their departure pushed me into a very dark place.

I was too scared to sleep in the hut alone, but there were always newcomers looking for a place to sleep, so I found some Afghans who seemed nice enough, and they were grateful to have the space.

Pustiwan tried to convince us to ride underneath some other trucks—those that had weighted suspension. When the truck is empty, some of the wheels fold up to save on tire wear. It's a good place to hide. But if the truck is loaded, you can be crushed to death in a moment. I knew someone who had died this way, and I didn't want it to be my fate—ground to bits in the suspension unit like lamb *kofte*.

Me and a few of the others refused. A couple of guys knew the risks and went anyway.

"Brave, no?" said Pustiwan.

"Foolish," I said, touching my cheek.

That Friday, Pustiwan appeared with a ladder under his arm. "A new plan," he proclaimed proudly, rubbing his belly.

We cut our way into a park full of trailers, ones waiting to be hitched to a cab and driven off.

Pustiwan tutted and paced back and forth, making a great show of inspecting each trailer, until he found one he judged to be just right. "Nice."

He placed his ladder against the frame of the canvas siding and climbed to the roof. He used a razor to neatly slice a hole large enough to fit a man. My face throbbed as he did it—I was not going first on this occasion.

"This is a good one," he declared. "Come, quick."

We climbed in one by one. The cargo was boxes of something heavy. I was relieved.

"Two days," Pustiwan said, his head poking through the hole. "Two days to England. Ha-ha-ha."

He spent the next ten minutes repairing his opening, then left without a word.

There were five of us. We sat for two days. No food or water. We lay in silence waiting and waiting. I dozed a little, but my dreams were surreal and disturbing—my blistered face hung broken from the wheels of a truck. I tried to lay still—to conserve energy, fluid, mental strength. After what felt like a week, the trailer shuddered into life as the cab hitched itself to us.

We were all suddenly wide awake.

"This is it," someone hissed through the gloom.

As depressed, angry, and exhausted as I felt, I got a rush of adrenaline. Perhaps this time I would make it. We began moving. Now came the sense of expectation. Either we would be making the long walk home, or we would be on a ferry bound for England.

I expected to turn left out of the park, toward the port. We'd made this attempt a thousand times. But, instead, the truck turned right.

"Are we going the right way?" I asked.

No one was sure. I could handle the hunger and thirst as long as I thought we were heading in the right direction, but we weren't. Panic set in. Someone started banging in the dark.

"Stop, stop. Let us out."

We all started shouting. I thumped my hand against a box—anything to get the driver's attention. Maybe he heard us, maybe he didn't. We kept driving. Further and further from the port.

After half a day the truck slowed to a halt. We heard voices outside. We were silent. The doors swung open, blinding us with daylight.

"*Halas*."

It was the police. We climbed down onto the roadside. I had no idea where we were. None of us did.

"England?" someone ventured.

"Belgium," the officer said. Then he burst out laughing.

The police tried to question us, but the language barrier made it impossible. In the end they made a great show of taking our names, then pointed to the nearest crossroads.

It was going to be a long walk.

WE WALKED FOR TWENTY-FOUR HOURS. MORE POLICE STOPPED us. We tried to explain our situation, but I'm not sure they understood or cared. "Calais," they said, pointing to a nearby road. We kept walking.

I don't know how I kept going. I just stared at my feet—left foot, right foot, left foot, right foot. I thought it would never end.

When we finally staggered into camp, I felt like the Jungle was the most beautiful place I had ever seen. I flopped onto my pallet bed and fell straight to sleep. No truck for us tonight.

At first I thought it was my mother, waking me for school—but the policeman was more persistent than my dreams. It was the weekly raid. I could have died. All I wanted to do was sleep. The horror of daily life now infiltrated my dreams. It was torture. The French police knew we didn't have any real options. All they wanted to do was humiliate us, take away what little dignity we had left. They wanted us to leave—to no longer be their problem—and so they left us no choice but to keep trying to go to Britain.

Each time I got caught, their taunts filled my ears: "No chance this time, maybe the next."

It was a perverse game, and I was trapped playing it for as long as it took—as long as I could keep going. It was the game without end. They thought it was funny. Every night was the same—cat and mouse with the police. Over and over.

Life only has value as long as you believe it is worth living. I was no longer sure. I was becoming detached from my surroundings. Nothing mattered anymore. The instinct to survive is strong, but when survival is all that there is, you are left with the obvious question: "Why go on?"

My life was a living nightmare. We kept trying trucks. The routine was the same. Break in, get caught, walk miles home.

One night the truck we were in took another right turn. I groaned inside. I no longer had the strength to fight it. The others banged and yelled, and then in desperation started to cut through the back of the trailer unit to escape. At that, the truck pulled over and the driver threw the doors open, glaring, baseball bat in his hands.

We had nothing to lose.

The six guys I was with leaped on the driver, kicking and beating him.

He dropped his bat and curled into a ball on the side of the road. "*Nein, nein. Halas . . . Bitte. Bitte.*"

I wanted to feel sorry for him. Part of me did. I hated violence. But ultimately I didn't care. He had a home, a family, a life. What did I have? Nothing. I was a dead man walking.

Passing cars honked their horns. The others stopped kicking and started running. We sprinted past a gas station with a car dealership attached. It had Mercedes flags flying from masts, along with red, yellow, and black ones.

"*Deutschland,*" someone puffed. "We're in Germany."

I was glad I was running so hard, because if I'd been standing still I think I would have fallen down crying. Not that I wanted to keep going. Only the Jungle waited for me.

I slowed until I was jogging by the roadside, watching the traffic rush past. It would have been so easy to step out. No more hunger. No more fear. No more Jungle. No more England.

I thought of my mother. I kept running.

Our little group split up at that point. I was exhausted and, try as I might, I couldn't keep going. Two men agreed to stay with me. We plodded in silence for hours. My reserves of strength—mental and physical—were all but gone. I wanted to lie down in the soft grass on the side of the road and sleep, and, if I was lucky, maybe never wake up. I fought these thoughts, but I was simply too weak.

Perhaps I would have given up but at that moment, out of nowhere, a blue sedan pulled over in front of us. A middle-aged brunette woman got out. She smiled. It was genuine, and I smiled back at the uninvited warmth she radiated.

"Refugee?" she said, in English.

It was a word I knew.

"*Oui,*" we said.

She opened the passenger door and I flopped gratefully inside. I don't remember her name. I'm not sure we even asked, but she was very friendly and clearly sympathized with our situation. She gave us a baguette and a bottle of water. I devoured the bread in seconds, washing it down with the water. Almost immediately, I began to feel a little better—I'm not sure whether it was the food, or just the simple act of caring she showed.

She dropped us off near the Jungle, where our makeshift shelter waited.

I should have been at rock bottom, but the act of kindness stayed with me. It sustained me psychologically for a few days—one small gesture from human to human. But the routine of life in the Jungle went on unchanged, and it wasn't long before I was lower than ever.

The burden was just too much. I was a young boy—I had no natural business being in such a place. I should have been in school, playing with friends, or spending time with my family; instead, I was constantly struggling to stay alive. Whether I actually wanted to, however, was increasingly on my mind. And I suppose that's why, sometime later, standing in front of a refrigerated truck full of bananas, I didn't hesitate. The six other guys I was with weren't so sure. "We'll freeze to death in there," one said.

I just kicked the mud off my worn-out boots and climbed up.

"Gulwali," they asked. "Are you sure?"

I shrugged. What difference did it make? Freeze in a banana truck, or freeze in the Jungle during this cold November? December would be even colder. Better to die than to go on living like this. I couldn't do it anymore—not another mouthful from the soup kitchen, not another mad sprint from the police or a driver, not another endless walk in the cold. My face was burned, blistered, and blackened—just like my soul.

If the other guys knew my state of mind, I doubt they would have followed. But they did.

"The kid knows what he's doing."

But I no longer knew anything.

CHAPTER 28

As I got in the refrigerator truck, I turned to the agent. "Don't say it."

Every time we got into a truck he waved and said, "*Khodahafiz. Farhda Englise,*" which means, "God be with you. Tomorrow, England."

He was Kurdish of course, but he said the words in Farsi. At first I found it funny; then it wasn't anymore. Now, it felt like he was taunting me, jinxing me.

We knew the distance from the truck park to the port entrance, we knew the sensation of going over the speed bumps that led to the port check-in, we knew how it felt when the truck slowed to join the queue for security checks. As each familiar sound or bump passed, I checked my mental map, plotting our course toward the elusive ferry.

The truck stopped, and I heard the French police open the door at the French checkpoint. I was convinced it was over. In my mind I was beginning to climb down and into their custody, just as I had done a hundred times before. But somehow, this time, they didn't see us. We were hidden right at the back, behind boxes of bananas, and perhaps that night the police couldn't be bothered really looking inside. They closed the doors.

I had entered uncharted territory. This was the closest I had ever reached after just over a month of trying—a month that had felt like three times as long.

Slowly, the truck drove forward to what I knew would now be the English checkpoints, and the doors swung open again. I flattened my body as if trying to becoming one with the boxes of bananas. Through a crack between the boxes I could see the guards pointing their flashlights around the dark spaces in between. Then their boot soles slapped hard against the concrete, and the doors slammed shut, plunging us into the comfort of absolute darkness.

The truck wobbled forward. My heart raced with each new noise. A hollow, metallic *clank-clank-clank* as each axle passed over a ramp sent a chorus of excitement through the trailer.

"We're getting on the ferry."

"Shhhh. We're not there yet."

I said nothing. I barely dared breathe. The truck went quiet as the driver turned off the engine, and for the next forty-five minutes the seven of us sat in absolute silence.

"Are we moving?" someone hissed.

I held my breath and concentrated. It was there—a discernible sway, a gentle rocking motion.

We were on our way to England.

I tried not to think of the deep black water sliding beneath us. This was a massive ferry—the biggest ship I had ever seen, let alone been on. I felt a little safer. These were professional sailors, not people smugglers with a leaking tub overflowing with pathetic, desperate men.

We began to relax and talk freely.

"Thanks be to God that he didn't turn the refrigerator on."

It was getting warm and stuffy in there, but at least it was tolerable. If he had turned it on, we could have frozen to death. There were many stories in the Jungle of people who had done so. That's why getting on to this truck had been such a risk.

Just a few hours earlier I had been ready to die. I was going to make it, or they'd find my blue corpse curled up on top of the banana boxes. Either way, I would be out of that living hell they called the Jungle in the port of Calais.

"Yeah," someone replied. "I really don't want to die, huddled shivering in your stinking arms."

We all laughed at that—refugee humor is dark. But without it we'd all have gone mad long ago.

"Hey," came a voice from the blackness. "George Bush, Tony Blair, and Hamid Karzai are all in hell."

"Is this a joke, or are you trying to cheer me up?"

We were really in a good mood now. It didn't quite feel real, and until I was safe and knew what my situation in England was, I couldn't relax. But at least, after months of trial and misery, I was on my way to Britain.

"Bush says to Satan, 'Hey, Satan, I need to make a phone call to Dick Cheney, to check how the war is going.' Satan replies, 'Sure, George—that'll be one billion dollars, please.' Bush isn't happy, but he pays the money and makes the call. Tony Blair then goes up to Satan. 'Hey, Satan, I need to call the queen in London—just to check on the war.' Satan says, 'Sure, Tony—no problem. That'll be one billion dollars, please.' Tony doesn't want to pay either, but what choice does he have? Finally, our glorious President Hamid Karzai goes up to Satan. 'Hey, Satan, I need to call Kabul to check on how the invasion is going.' Satan says, 'No problem, Hamid. That'll be fifty cents, please.' When Hamid goes to use the phone, George Bush and Tony Blair rush up to Satan: 'Satan, Satan—why did you charge us a billion dollars, and Karzai only fifty cents? It's not fair.' Satan turns to George and Tony: 'Look, guys, it's totally fair. You're phoning Washington and London—that is long distance. Hamid's is only a local call.'"

I laughed and laughed. We all did. It felt good. My face itched a lot in the still air of the truck, as the burns were far from healed. I took my cream from my bag and smeared it on. It offered some relief.

After an hour or so, the truck burst into life. The ferry engine changed note.

We started to talk more loudly then, because we had made it. It didn't matter if we got caught—we had made it to England.

WE'D BEEN DRIVING FOR AN HOUR ON THE ENGLISH SIDE WHEN a fight broke out. The other six guys were arguing about whether we should try to open the door from the inside when the truck stopped, or try to get the attention of the driver now by making a lot of noise by banging or trying to move the boxes around. The men knew they had to run for it as soon as they could, because otherwise they would be arrested and feared they'd be deported.

I was tired of running. I just wanted to wait until the doors were opened by the driver. I was so exhausted and sick I couldn't run anywhere. I didn't know if I'd be allowed to stay or not. I also worried I might be arrested and thrown in jail like I had been elsewhere.

But I didn't think the UK would be like Iran, Bulgaria, or even France. I knew enough about it to know there wouldn't be a policeman with horsewhips or kidnappers waiting to grab us, so how bad could anything else be?

The others weren't happy, though.

"What do you care, kid? You'll be fine. Don't be selfish. You know they will take care of you. But what about us?"

"You run. I will stay here. I don't care," I said.

In the end, we agreed to try knocking but the driver couldn't hear us. Eventually, when he stopped at his destination, he opened the door. The others were poised and ready to jump but the driver saw us, a look of total shock on his face. But despite his shock he was fast, and slammed the doors shut before anybody could get out.

"No!" someone cried.

I had had my first glimpse of England, but now I was locked up in the dark again. We were scared. We didn't know what was going to happen now.

After a quarter of an hour, the doors swung wide open. Four police in blue uniforms stood there.

They ordered us out. I blinked in the bright daylight. It all looked depressingly familiar. It was yet another warehouse complex, with parking for dozens of trucks.

I smelled British air for the first time in my life. It didn't smell of fish and chips or roast beef—as had been the joke among the Jungle refugees. But it wasn't bad, either. Maybe it smelled of freedom. The weather was good—surprisingly sunny for November, and much warmer than the day I had left France. Like everybody, I had heard it was always raining in England.

We stood in a huddle as the police questioned the driver. I think he was Dutch. They wanted to see his license and they seemed to think he must have known we were in there. He was sweating and nervous.

I felt bad for him and wanted to tell the police he didn't know we were in his truck, but I kept quiet. Eventually they loaded us into a police van and took us to a police station near Dartford, in Kent.

There were Pashtu translators there, to help the police with the questioning. We were under arrest, they explained. I had my fingerprints taken. I'd been arrested before, but never so politely—they explained that I was being arrested and told me what my rights were; they said I could stay silent and I didn't have to say anything unless I wanted to. It was as though they were asking my permission.

I hated the police by now, after all my experiences; but this felt different. I was unsure of what would happen to me but I was no longer afraid. I had seen it all, suffered it all already. What would these well-mannered English police do to me that the Iranian, Turkish, Bulgarian, or French police had not done already?

I told them my age. "Thirteen."

They laughed at me—humiliating, bitter, cynical laughter.

They did not believe me, even after the translator repeated the question several times.

After interrogation, they put us in small cells for twenty-four hours. It was the longest twenty-four hours ever. I was thinking, "Okay, I am in England—but what now? What if they deport me? Have I made it this far only to go all the way back to the start?"

Before, I hadn't thought that they would; now I wasn't sure. Being locked up felt like Turkey, like Istanbul, back in the same police station. The food was horrible, and all of my anxieties came flooding back.

The following day they put us into a minibus. They didn't tell us where we were going. Everything was unclear again. I began to really think the worst when I realized they were taking us back to Dover. I could see ferries like the one I had just crossed on, cutting their way through gray wintry waves toward the distant French shoreline.

We were led into a building with the sign "Immigration Removal Center" over its main door—one of the men I was with explained to me what this meant. I was sure they would send us all back to France, or even Afghanistan—that made me scared like never before. I did not want to go that way. Going back now would be worse than death.

"Please, God. *No,*" I prayed.

At the vast refugee center we were put in a featureless waiting room. It felt familiar, like so many places I'd seen before, filled as usual with all the faces of the world. We started talking to people, trying to get information about what was happening. Some were there to sign paperwork before being deported. They were fingerprinted as they left, looking dejected and broken. Many of the others had just been arrested like us.

The atmosphere was distressing. Some people were crying. Others were just sitting huddled, their arms wrapped around themselves, as they rocked back and forth.

To have come all this way and to have fallen and failed at the last hurdle was just too much.

I shared their feelings—I was filled with dread. Clearly, England wasn't the welcoming place I'd been led to believe it was.

I was taken alone into a room and questioned by two immigration officers. With a translator, they interrogated me for hours. They kept asking about my face and why it was so badly burned. They insisted on looking at my tube of burn cream, and wanted to know exactly what it was. I was sent to see a doctor, who examined both my face and the suspect cream.

I talked to many different people, all government officials, and all through a translator. They all asked the same thing: "Why did you come? How did you get here? When did you arrive?"

All day I was there. They asked detailed questions about my family and if I wanted to claim asylum. I told them I had left because my life had been at risk, so yes, I wanted to claim asylum.

It was the first time on my whole journey I had been asked about asylum.

It was strange. The question itself didn't feel comforting—not like I thought it might have. I had expected it to be like being handed a prize at the end of a race, but it didn't feel like that. I was in pain, I was tired, and I was bewildered. I knew the process had to be done, but I was too traumatized to answer their questions properly. I had been awake for days and I hadn't had any food for the same length of time.

At least the immigration officials, when I told them my age, just wrote it down and didn't laugh in my face, like the police had. I asked one man about my brother and whether they could find him for me. He did laugh at that, and told me there were sixty-five million people in the country—so, no, they couldn't just find him. He didn't even write down his name.

That made me angry. I had seen some people being collected by relatives, so I knew that it must be possible to try to find people. And if every migrant or refugee who arrived was registered in a computer system, why could they not at least try to search for his name for me to see if it came up?

I seemed to spend hours in that waiting room. There was a coffee machine in one corner that squirted black liquid into little brown plastic cups. It was hot, there was plenty of sugar, and it was completely free. I must have had eight cups and it made my head feel a bit funny. But it still didn't make up for the lack of food—I was starving, and no one had offered me anything to eat.

Eventually, someone came to tell me that I was being assigned a social worker, and registered as an asylum seeker. I felt some relief—at least I wasn't being sent home straight away. An official gave me my ID card showing my name and date of birth, and a document showing my illegal entry status in Britain. I was told I had to return to the center regularly and report in.

Seeing my picture on that ID was strange. There was little sign of the fresh-faced boy I had been just one year before. My face was scarred and pockmarked, and I was thin and drawn.

But I was proud—I had done it. I was here, and they weren't sending me back—at least, not yet.

Later the same afternoon, a social worker took me to a hotel. I didn't see the others from the banana truck again. I still don't know what happened to them. The hotel was on old Victorian seafront building. It was nice, clean. The hotel manager showed me to a small single room and told me to take a shower. I didn't need to be asked twice.

As I walked down the hall toward the communal bathroom, carrying a towel, I spotted a pair of shoes neatly placed outside a door. My heart leaped. When you've used someone's feet for a pillow, you don't forget their shoes.

I banged on the door. Qamandin opened it. His face broke into a familiar smile. We hugged. It was so good to see him—and in England too. He told me Jan had been separated from him in the refugee center, and he didn't know what the authorities had done with him.

That night, in the hotel, we had long, hot showers and then a big supper of chili con carne and rice. I slept like a baby that night.

IN THE MORNING, I FELT LIKE I HAD BEEN BORN AGAIN.

"Let's go and see England," Qamandin said, wiping the debris of his breakfast from his mouth with a paper napkin, like a proper English gentleman.

We walked into Dover town center, following the seafront. The clean sea air felt great on my face—cooling, healing.

We climbed up the chalky cliffs toward the black stone castle that sits high above the town. From our vantage point I could see dozens of ships crisscrossing the English Channel.

It was very strange to see Calais on the horizon and know how much pain there was over there, in the Jungle. It was getting dark now, and I knew that not so far away, thousands of men, women, and children would be getting ready to begin their futile, nightly task of trying to cross this unimpressive stretch of water. A shower, clean clothes, freedom, and a place to stay. That was all I had, and I felt like a king. Why couldn't everyone have access to something so simple? Why were human beings given as little value as the fleas that bred in the makeshift tents that I had, less than forty-eight hours ago, called home?

Mentally, though, I was still a mess. I still felt completely alone and depressed. I needed my family—I needed to find Hazrat. There was only one way I knew to do that.

And so, once again, I ran away.

I CHANGED THE LAST OF MY MONEY—ABOUT 100 EUROS—INTO British pounds. I found my way to Dover train station and bought a train ticket for London. I didn't know where Hazrat was, but I needed to start my search somewhere.

As the train pulled out, I felt quietly confident. I had managed to work my way around Rome, Athens, and Paris and find people I had a connection to. London couldn't be any different.

When I got to London Bridge station, I had second thoughts. London seemed so impersonal, so big and confusing. But I applied the formula I had used in every other European city I'd passed through: I walked the streets around the station and searched out Afghans, and asked them if they had seen Hazrat Passarlay. They all said the same thing—go to social services.

This made me realize how stupid I had been. There was a system here—I needed to learn how it worked. Walking the streets talking to Afghans was fine when I was trying to locate an agent or a smuggler, but here there was a different way of doing things.

It was time to go back to Dover.

THE NEXT DAY, A SOCIAL WORKER CAME TO SEE ME. SHE WAS Eastern European. I hoped the fact that she was a refugee too might make her sympathetic to my situation.

I only knew a few phrases in English, basic questions such as, "What is your name?"—the French police had generally spoken in English to me when they arrested me, and I found the language much easier to pick up than Greek, Italian, or French. But her English was not much better than mine. I also found her very aggressive and abrupt. She refused to believe I was thirteen, and without proper evidence there was little I could do to persuade her otherwise. She looked at me like I was something distasteful stuck to the bottom of her shoe. She told me she was from Kent Social Services, and that I needed to be interviewed so they could decide what to do with me. I would also be required to undertake an age-assessment test.

I spent the next few days waiting and worrying, spending time with Qamandin. Each day at the hotel people were interviewed, and often taken away afterward. Qamandin told me that some went to Appledore, a special unit for refugees deemed to be young enough to be registered as unaccompanied children; other people were put in touch with relatives and allowed to go and stay with them. I wondered what would happen to me.

I was still terrified that I would be deported, and wondered every day what Hazrat was doing.

ON THE DAY OF THE AGE ASSESSMENT, MY SOCIAL WORKER came to collect me. She took me to an office at Kent Social Services. I sat around a table with five different people—some were from social services, some were from Kent County Council, and one was a Pashtu-speaking translator.

They quizzed me for hours. None of them smiled. It was worse than the immigration center.

Some questions were complete nonsense to me, such as asking me to name streets in Afghanistan. It was almost as if they didn't believe I was Afghan. They asked me the name of the main square in Jalalabad and what famous rivers in Afghanistan were called. I gave them the answers they wanted.

Then the translator explained to me that they felt I was too clever and smart to be thirteen years old—as if correctly answering their stupid questions was evidence of my adulthood.

At the end of the meeting I was sent to a waiting room. After a long while, they called me back. A man with a face that looked as worn and gray as his suit, one of the council people, spoke, and the translator outlined what he was saying.

"Based on this interview, we come to the conclusion that you are sixteen and a half years of age. Therefore your date of birth is the first of May, 1991."

This was crazy, beyond the realms of make-believe. After all I had been through—the cruelty, mistreatment, and abuse—to be assigned a new birth date by a committee stranger was more than I could take.

They gave me an age-assessment document with their findings, and my new date of birth on it. I couldn't read what it said, but I could make out the date of birth.

I tore it apart and threw it on the table. "Thank you very much. I don't want this. You keep it," I said, scattering the confetti in front of them. The translator relayed what I'd said in English but my actions made it obvious. They all looked completely stunned.

I may have looked older than my years, but what had I just gone through? My face had been very damaged by the burns, giving me a much older appearance and, besides, I had had to grow up fast and learn to act tough. How could these people not understand any of this? Had they not seen the record of my medical examination on arrival at Dover? Did they have no concept of what living outdoors in all weathers on a near-starvation diet might do to a person?

I stormed out in protest. I had no proof to dispute their absurd findings, and common sense wasn't a language these people spoke. I felt that instead of smugglers and agents, I was now in the hands of new strangers, and these bureaucrats were now in control of my life.

To be treated like a liar by this committee of officials was truly soul-destroying. After my outburst, I refused to say another word. In the face of such stupidity, what did my cooperation count for, anyway?

A few days later, some officials arrived at the hotel.

"Gulwali," said the manager, two unfamiliar faces standing behind him. "These men have come to take you to Appledore."

CHAPTER 29

Appledore was a large hostel for migrant and refugee minors who had no family. It had twenty-nine shared bedrooms and a big communal kitchen and sitting room. There were around twenty or so under-sixteens in the building at the time I was there. The majority were Afghans, but there were also Africans and Arabs. All of them, like me, had traveled alone.

We had care workers looking after us. They were very nice people. My key worker was Scott, the assistant manager of the center. He was very young and had the ability to make me laugh, but he was so busy it was really hard to get any one-on-one time with him.

It was a very well-equipped place. The rooms were clean and tidy, with modern furniture. There was a table-tennis set and a football pitch outside. They really did try their best for us.

We were also expected to use the time to learn more about the UK. We had classes about British life: its history, monarchy, culture. One video we watched told us the UK was the father of democracy and a big leader in human rights.

We also had life-skills classes, such as time and money management, and how to use public transport. We won brownie points for cleaning our rooms and helping the staff in the kitchen. We used to go shopping for the weekly kitchen needs—the idea was that by doing so we learned how to manage and allocate the budget accordingly.

I still preferred the company of older people—maybe my time spent with Baryalai and my other older friends en route had made me this way—so for me, hanging around the staff and offering to stack the dishwasher, unpack boxes of supplies, or clean the shelves was preferable to talking to the other depressed, sad children like me. I didn't want to be reminded of my own story, let alone hear someone else's. I couldn't cope with it.

We were given a budget for clothes and had to write out lists about what would be appropriate for a particular location or weather. They needed to see that we could work out prices and decide what was a good buy or a waste of money. It was all designed to ensure we could cope on our own when we left the center to live independently. I think it was really useful.

The first time I went shopping was really fun. I had been given a clothes allowance from my social worker—I think it was about £100. One of the care workers, Lorraine, took me shopping and helped me choose lots of things. She took me to a very big store she knew about, which she said sold bargain goods at the very best prices. She said it was where all sensible English people shopped, but that rich people laughed at these kinds of places.

Inside, I was overwhelmed by all the choices—the shop sold everything, from household cleaning supplies to toys and clothes and bed sheets. We had so much fun as Lorraine helped me cruise the aisles finding discounted items and special offers. We managed to get so much—socks, towels, shirts, and jeans. I'd never really had more than one change of clothes, at least since I'd left Afghanistan.

When Lorraine and I got back to the center, we were both really proud of how much we'd managed to get for our money. I was so grateful to her. It gave me a real sense of achievement—the first I had had since getting here.

I HAD BEEN THERE FOR JUST OVER TWO WEEKS WHEN Qamandin was sent on from the Dover hotel to Appledore too. It was great to be allowed to share a room with him. He had been through a similar age-assessment process, but he had a different social worker to me, and different people assessing him. They believed him when he said he was fifteen. Eventually, he was sent to a foster family in a town called Hastings, leaving me feeling very alone. Why could I not be fostered too? I was still deeply unhappy about my own age assessment and dispute, but no one was willing to help me challenge it.

One of the things I liked best during my time at Appledore was that every Friday they dropped us at the train station and let us catch a train to Tunbridge Wells, to a local mosque. Meeting other Muslims was comforting . . . But Tunbridge Wells didn't seem like a place with very many immigrants: walking around there, where everyone looked very English, made me feel quite self-conscious. I felt that everyone was staring and looking down on me. But it was a really pretty town, and I enjoyed looking at the architecture.

Letting us go there on our own was a big deal for the staff—they had to trust us. Perhaps once or twice I considered running away, but I didn't because I really thought that by staying at Appledore I could sort out the age problem with social services, and begin the process of finding my brother. I also liked that I knew my way to the mosque and past a few of the nice sights, and I enjoyed the feeling of proudly showing off my newfound knowledge to some of the other new arrivals.

We even got taken to London, to the National Portrait Gallery. Art was a new thing to me. Images are not Islamic, and photography was banned by the Taliban, so portraits and pictures weren't something I had seen much of before. I found it a bit strange staring at the faces of people I was told had died hundreds of years before; to be honest, I found it a bit boring. I think, even despite it all, there was a little bit of me at that point that was a typical teenager. I did like the British Museum, though.

Every week we were given £10 pocket money by the care workers, which most people used to call their families. I was so sad that I couldn't do this. We also did community work, helping to clean a nearby steam train center, and helping Kent County Council environmental agency with planting trees. This I loved. It reminded me of being a child again and cutting branches for the classroom roof. It was good, physical work. I know for sure today that there are fifteen trees in Kent that I planted, and I'm proud of this.

A MONTH INTO MY STAY THERE, AND I WAS BEGINNING TO wonder if I would ever leave Appledore. Once again, my mental health was not good—I was feeling isolated, alone, and numb.

One day, I was sitting in the kitchen chatting to a boy called Kiran, a new arrival from Balochistan in Pakistan. He told me he had been arrested by the Pakistani military and hung from his feet and beaten with small canes. He said he didn't know why this had happened because no one in his family was involved in the Baloch independence movement. He said they had no interest in politics at all—very unlike mine.

I wanted to support Kiran and show that I cared, but I didn't honestly have the ability or energy to comfort him. I was just wondering how I could walk away without being rude when I heard a voice behind me.

"Gulwali."

It was Shafique. I leaped from my chair so fast I nearly knocked it over.

Shafique hugged me and whooped so loudly that some residents came to see what was going on. Once we'd calmed down and talked, he told me his story. He'd managed to get across the English Channel from the French port of Dunkirk by hiding in some kind of metal storage box attached to the side of a truck. He said he had known it was very dangerous because there would be limited oxygen in the tiny space but, like me, he had reached such a point of desperation by then that he hadn't cared if he lived or died.

I was so happy he'd made it and didn't have to spend the rest of the winter freezing in France. I was still in the grip of despair but having him there helped so much—just knowing there was someone there that I didn't have to explain myself to.

Not long after this, I was told I would have a Home Office interview. I was assigned a legal representative, who came to see me at Appledore in advance of the interview, to go through my case. She seemed nice, and didn't argue when I told her how old I was. She seemed to believe me, and so I hoped she'd be able to do something. I also begged her to help me find Hazrat. She said that I needed to talk to the Home Office about this myself, but that she would put a note about it in my file.

By this stage, however, I was so angry and untrusting that I didn't tell her as many details about what had happened to me in Afghanistan as I should have done. Everyone kept asking me the same questions all the time, but it felt to me that no one was really listening, so I think I had just given up. I didn't have the energy to keep going over it.

I was not an adult, and I kept telling them I wasn't, but they were treating me like one. My feelings intensified when I realized that the Home Office had given me an adult interview and not a child's one—one where I would have been treated with more gentleness—because the social services had designated me an adult. I would eventually discover too, that though the Dover Immigration Removal Center had believed my age in the first place, they had since changed it to be in line with the social services' assessments.

The day I went for my interview I felt sick inside—it took every ounce of motivation I had to make myself go. My feeling was that they wouldn't believe me, so what was the point?

I traveled to a place called Croydon in London for the interview. One of the care workers came with me. Croydon itself felt gray, impersonal, and ugly—matching my mood completely. We arrived at a place called Lunar House. It was even grayer—there were queues snaking out of the door, while the scene inside set the tone: there were more long queues, and the sounds of people sobbing or arguing.

I was dismayed to discover my legal representative wasn't there, as she had promised me she would be. One of her colleagues came instead, but he knew hardly anything about me; it was obvious he had only read my notes ten minutes before.

I was taken to a room and, through a translator, I was grilled as though I were on trial.

If I had thought the social services' assessment was bad, this was a whole different level of awful. It was as if they were deliberately trying to make me feel guilty and humiliated for having dared to come to the UK. I felt victimized and criminalized.

I tried to tell them that I was only thirteen, but a man at the table said, "Sorry, we are not here to discuss your age today. That has already been done and dealt with." They then told me that the information I was telling them was new and asked why I hadn't told them these things in my witness statement—a statement I had made for my solicitors. They also said there were differences in the account I had made when I spoke to the immigration center in Dover.

Maybe there were, but it can be hardly surprising, surely. I had been traumatized: I had just got off a refrigerator truck after a long journey, I hadn't eaten, and I was in shock. I didn't know then how the system worked, or that you were supposed to remember every key detail at the first interview lest you pay the price for it later. Couldn't they see that if I had made mistakes, they were genuine ones?

I just couldn't speak anymore. I started crying, which made me feel even more humiliated. Then one of the men told me to stop acting.

At that I lost it. I started hitting the table, banging and banging it with my fist. No one said anything, they all just stared at me, as if this was also a sign of my criminality.

The whole process was so impersonal, so strange. Telling my life story to strangers was utterly at odds with my culture: we respected each other's privacy—it wasn't even something I had done during my time on the road. Now, here I was having to tell it to a bunch of strangers sitting around a table.

And it wasn't that I didn't want to tell anyone, but that I needed to feel safe and secure to be able to talk. This process was the opposite of that.

At the end of the four-hour interview, they told me they would make a decision on my application for asylum and that I would be informed.

I already knew it was over.

BACK AT THE CENTER I WAS SO DEPRESSED I COULDN'T FUNCtion. But more bad news was to come: Shafique had had his age assessment, and they had decided he was eighteen, when in fact he was sixteen. Immediately after his interview he was taken by immigration officials to a detention center. We didn't even have time to say good-bye to each other.

He remained there for a week and had been told that he would be deported to Afghanistan within thirty days, but he got lucky in the end. He had a very good legal advisor, who fought Kent County Council about his age and the Home Office about his detention. The lawyer arranged for a doctor to examine him to reassess his age. The doctor accepted he was just under sixteen, and he was released back to Appledore. He was granted two years' leave to remain in the UK and placed with a foster family in Gillingham.

I was able to visit Shafique there once he got settled, but I don't think he liked being there. The family was kind but he and I both got the sense they were doing it for the money, not out of love. Seeing this made me think that maybe it wasn't so bad I hadn't been fostered after all.

My depression could not have been worse at this point. I recalled the children's home in Italy and the lovely kind Alexandria, and I began to wish I had stayed there. At least they had treated me more kindly.

Added to this, I despaired of ever finding my brother. No one would help me, and in a country of so many people, how did I even begin?

One day I got so angry I walked out of the center in protest. I made my way to the station in the late afternoon; it was late November, so it was dark when I got there. I got a train to Tunbridge Wells and just walked around. I fully expected to be arrested. I think I had lost my mind. I was like a zombie, suicidal, walking in the middle of the road, hoping a car would hit me.

After a few hours of walking around like this, I realized I had nowhere to go. I had to go back to Appledore. I called the center to tell them I was at the station. One of the care workers answered the phone. If I had wanted understanding I didn't get it.

"Gulwali. I am disgusted at you. We have done so much for you—why are you so ungrateful? We were worried about you."

When I got back I ran to my room. I couldn't look anyone in the eye. I felt like I had done something very wrong and felt so guilty. I thought they would all hate me.

"Gulwali, we have lost all respect for you. You have lost the freedoms and privileges you had. Now you are not allowed to leave the center."

I understood my actions were wrong, but couldn't they see what a mess my mind was in?

The nightmares I had had in Italy and Greece had returned. I would wake screaming, thinking that my bed was on fire, the burning sensation so real that I would run into the bathroom to douse myself in water. Sometimes I got so panicky I couldn't breathe or move; I could only lie on my bed willing myself to breathe and calm down. I missed my mother and grandmother so much. I longed for one of them to be there to comfort me in my night terrors, to hold me and tell me it was okay, that it was only a bad dream. Instead, I had to tell myself that. I was my only comfort.

A couple of days later, the others went on a day trip to London to go on the London Eye. I wasn't allowed to go as punishment for my escapade. I accepted why I couldn't go, but it made me feel even worse.

For the first time, I felt really suicidal. At Calais, I hadn't cared whether I lived. Now I only wanted to die.

I was in my room when they all came back.

"Gulwali, open the door." One of the other residents was banging on my bedroom door. I ignored him—the last thing I wanted to hear was about their fun trip. "Gulwali. We need to tell you something."

"Go away, leave me alone."

He knocked again.

"Gulwali," he said. "We found Hazrat."

CHAPTER 30

THE NEXT DAY MY DREAMS CAME TRUE. I WAS SHAKING AS the care worker handed me the phone. I couldn't breathe.

"Gulwali?"

I started to sob. After that, it became a blur. Hazrat was almost hysterical with joy, babbling and blurting out his story. He told me he'd never given up hope and had shown my passport picture, the one the agent had given him in Peshawar, to everyone he met on the road.

He had arrived in England six months ago, and was living near Manchester. He was taking English language lessons at college and working in a shop. He was living with a friend, and had been granted two years' leave to remain as an asylum seeker.

Neither of us could believe the coincidence of how he'd found me. He had been in London on a college trip, walking along the South Bank. He and his friends had seen some fellow Afghans by the London Eye—my friends from Appledore. They had chatted and, during the conversation, one of the Appledore people had said there was a boy in the center who looked just like him. Hazrat had asked the name of the boy and when they told him he began screaming, "That's my little brother. That's my little brother."

His social worker arranged for him to visit me a few days later. I was so nervous. Would my brother be the same person? I wondered if the indignities and traumas we had both suffered had changed him.

As it was, I hardly recognized him. He was so grown up. As he wrapped me in his arms, I couldn't help tease my big brother, just as I had always done: "Hey, you got fat. How did that happen?"

We spent the day together sharing stories of our journeys. His had been very similar to mine, with lots of going backward and forward, but he had spent less time in Istanbul and more in Greece. We also cried together as we thought of our family back in Afghanistan.

After he left, I thought my heart was going to break all over again. All I could think of now was two things: getting out of Appledore to be with him, and convincing the Home Office I was only thirteen. But bad news was to come.

I had been refused asylum.

I had initially been granted discretionary leave to remain for a year, starting from my arrival date of November 17, 2007, due to expire in November 2008. Then, as a result of the incorrect birth date they had given me, they told me I would be seventeen and therefore I would have to leave the country or be deported.

I was relieved to be able to stay for a year at least—it was better than nothing—but I was still so very angry that the age that was being forced on me was still the same.

And it had got worse—the latest documents given to me stated: "Nationality Unknown."

How could they say that?

The worst blow came when I was refused permission to stay with Hazrat, and told I had to stay in Kent.

The one piece of good news I received at this time was in March 2008, when I was given a new social worker. She was Iranian by descent, so we could communicate in Farsi as well as in my broken English. That helped. I also felt culturally comfortable with her. I felt the social worker before her may have been a bit racist, and that was why she had been so abrupt with me.

Nassi was more supportive, but even she said my situation wasn't good and that I would probably be deported.

Through her, social services found me a flat in Gravesend in an accommodation block with other asylum seekers. Mine was nice: a modern two-bedroom flat. The other occupant was also Afghan, in his late twenties, but he seemed to see me as an annoyance and pretty much ignored me. It was noisy. Some of the others played music all night long, which only intensified my loneliness. So did the nightmares.

My flatmate thought my nighttime screaming was deliberate and complained about me to Nassi. The humiliation of being told off for something I couldn't help was the final straw; I had reached my limit.

The next day I calmly walked into a pharmacy and bought a bottle of paracetamol.

Then I swallowed them all.

CHAPTER 31

My flatmate found me unconscious and called an ambulance. At the hospital they gave me something to make me vomit up the pills. My stomach, which had suffered so much already, was in agony as I heaved and retched time and again, vomiting up green bile into a bucket. The hospital staff were kind to me, and I appreciated it.

I stayed there a few days before being sent back home on my own. My lonely life continued.

I had no schooling, no work. Days just ran into endless, boring, lonely days. I bought a few books from the local charity shop and used the time to try to improve my English. I also read the Quran, something that always gave me comfort. I was in terrible inner turmoil. Suicide is expressly forbidden in Islam. I felt very guilty for what I had tried to do. Yet I had a strong feeling this had been another test of my faith. I felt suspended, floating between the brutal earthly reality of my existence and the tranquility of my prayers. What I had done was wrong and I knew it was wrong, but at the same time I didn't feel that God had abandoned me.

Nassi got me into North West Kent College on a part-time basis to study English, two days a week. I showed the teachers my first ID, which said I was thirteen, telling them I shouldn't be there and that I should be in school instead. But again, they said they couldn't help me.

I attended the college for a couple of months. The other students were, of course, older—all college age—but they too were struggling to adapt to life there. A couple of them were drinking heavily to quell the boredom, or maybe it was a way of self-medication to numb the trauma. One stopped coming to college suddenly, and we heard he'd been arrested for assaulting someone.

I think that after having grown up in a rigid cultural system like mine, it can be very hard for new arrivals to cope with the sudden social freedoms that Britain offers. And with no money, it's hard to keep busy. I received a young person's allowance of £94 per fortnight, with which I had to buy everything I needed, including bus fares. I was good at budgeting and I had learned to be a bargain hunter by seeking out special offers and discounted food in the supermarket, but it was still a struggle.

Shafique studied with me at the same college, and came over to visit me often, and that helped, but even his best efforts weren't enough. I missed my brother so much. Being in the same country as him after all this time but not being near him was torture.

I used all of my allowance and bought myself a train ticket to Manchester. Once there, I called my social worker, Nassi, and said I wasn't coming back.

"You have to. You are registered in Kent. This will cause big problems for you. Go and visit your brother by all means, but if you don't come back I will report you as missing."

I couldn't see how it could get worse.

I DECIDED THAT IF NO ONE WAS GOING TO HELP ME, THEN I had to help myself. In Manchester, I literally walked into a couple of local schools and asked them to accept me. They said their hands were tied and they couldn't. Once again, I realized that this country has systems: systems I needed to understand before I could challenge them.

Proving my date of birth was the first step.

Hazrat and I went to the Afghan embassy in London. I applied for a new passport, giving them my immigration ID card as evidence—the very first one, which had my real age on it. I also had to get letters from two other people who could verify who I was and how old I was.

When I got the passport, I sent it to the Home Office as proof, but my plan backfired. The Home Office asked, if I was an asylum seeker, why was I going to my national embassy? But that was a silly question in my case—I hadn't been persecuted by my government (as many refugees from other places have been, causing them to leave their country and ask for asylum elsewhere). In my case, it had been a combination of the Taliban and the American military who had forced me to flee.

I was also in a lot of internal conflict over my views toward the UK. I very much wanted to stay now that I was here with my family, but I saw this country as partly responsible for the war in my country. Britain was the key U.S. ally. I also believed that Britain was responsible for a historical chain of events that plunged Afghanistan into a state of instability and chaos today, going back to the Durand Line and the two Anglo-Afghan wars in 1839 and 1868. I felt a confusing mixture of gratitude and, if I am honest, hatred. I couldn't handle the conflicting emotions.

While I was in Manchester, some of my brother's friends told me about a place called Starting Point, in nearby Bolton. It was a special educational center for children who had just arrived in the UK. Starting Point offered intensive education in basic subjects, as well as a strong program in English language support.

The head teacher, a woman called Katy Kellett, agreed to see me. As I sat in her office drinking a hot chocolate, a lovely warm feeling of calm gushed over me. She made me feel instantly safe as I explained to her my situation, making sure to stress the age dispute.

The next words she said changed everything for me: "Gulwali, I believe you. You can come here to study."

She told me age disputes like mine happened all the time. I could barely contain my joy as she asked the secretary to print me a letter confirming my place there, and get me a new uniform from the school shop. Mrs. Kellett also promised she would contact Nassi and sort everything out. Her kindness was overwhelming. She then spoke to Manchester Social Services on my behalf and asked them to transfer my care from Kent to them. But they said they couldn't, telling her that they had had bad experiences of dealing with Kent in the past and wouldn't get involved. Eventually, I was allowed to stay in Manchester, but I was still officially under the care of Kent—it was very confusing.

But Mrs. Kellett was successful in persuading Kent Social Services to get me a place to stay in Manchester. Hazrat had been living with friends, and they had got sick of me being there too. She arranged for me to stay in a place called Bedspace in Hulme, in the south of Manchester. It was similar to the first independent living accommodation in Gravesend—shared flats in a block, and this time my flatmate was a Kashmiri man in his thirties. Because the Home Office was still insisting I was sixteen and not thirteen, I had to be housed with adults. I had to catch two buses to get to Starting Point each day, but that didn't bother me. I was just so happy to be able to go there.

THE YEAR 2008 CAME TO AN END, AND WITH IT PASSED MY fourteenth birthday. For the first time in two years, I was beginning to feel like a child again. Starting Point made me school captain, which made me so proud.

Being in school was really helping to soften my harsher views toward the UK. I loved that here I was valued and that people cared about my achievements. Every time I got a good mark or a gold star a little bit of the normal child in me came back to life.

The other children there were from all over the world: some were asylum seekers, others were from refugee families who had moved to the UK to work. The numbers attending the place fluctuated: sometimes it was fifty, sometimes as few as ten. I truly appreciated that Britain was willing to give children like us a chance. Starting Point became my family. I loved sitting at the front of Mrs. Brodie's class, raising my hand to ask questions.

"Gulwali, can you keep quiet for just two minutes and let the other children ask a question for once," she teased, smiling at my enthusiasm. But I was so determined to learn and to soak up every minute. During break times she used to make me toast and tea and sit and chat with me. I loved Mrs. Brodie and Mrs. Kellett so much—the two of them did so much for me.

Once a week, Starting Point pupils visited a local old people's home. I loved it because I loved elderly people—they reminded me of my grandparents. I felt so sad for these old people who had to live there, away from their families. Some had no family at all.

I think, during this time, I was overwhelmed by it all. After all the battles to survive, I couldn't quite cope with the fact that my life was now improving. School was about trying to reclaim a sense of normalcy, of trying to be a child again.

But inside I was terrified, tearful, or angry by turns. My mood swings were terrible. I scared myself at times with my rages, but I had no idea how to deal with them or who to ask for help. I didn't want to confide in my teachers how angry I felt because I didn't want them to think badly of me.

And every night I woke in sweats and tears as the nightmares gripped my soul.

The more normal my life became, the less I felt in control. The more stable the circumstances around me, the more trauma I felt. That confusion left me crippled with depression and fear. But I had no one to turn to.

After my suicide attempt, I had been offered counseling. It helped a bit but I only had a couple of sessions. Because of the age dispute with social services, I didn't fall under child protection services so I wasn't really monitored.

I did think about telling Mrs. Brodie and Mrs. Kellett how scared I was but stupidly I thought they'd dislike me if I told them the truth. I couldn't see a way to escape my own mind. I started to plan my own death. I thought about jumping off a nearby motorway bridge or maybe throwing myself under a train, but then I thought it might hurt too much so I got scared.

Instead I went to the chemist and bought as many pills as my allowance could afford. I swallowed the lot.

THIS TIME I VERY NEARLY SUCCEEDED IN DYING; ENGINEER found me because he came round to visit unexpectedly. He rang 999 and I was rushed to the hospital in an ambulance. Engineer came with me and called Hazrat to tell him.

Hazrat cried by my bedside. He made me swear I would never do something so silly again. "Did we go through all this just so you could die on me, Gulwali? What will I do if I lose you?"

I felt guilty, of course I did. But I was still so lost and unhappy, I probably would have tried to kill myself a third time. But my mother's voice saved me. One of Hazrat's friends had been deported back to Afghanistan, and promised to make contact with our family. I didn't really expect him to. But three weeks later he called Hazrat with a mobile phone number for her. He said she had been overjoyed to have confirmation we were safe. Qubat had told her we'd reached Greece, as far as she had paid us to get. But without hearing our voices for over two years, she had no real idea if Qubat was lying or not.

Within seconds of receiving the number Hazrat was dialing it.

He put it on speaker. "Boys? Is that you?"

Her voice was nervous, trembling. I could feel myself shaking.

"Morya, it's me, Gulwali."

Then suddenly we were all laughing and crying at once. She told us how the little ones were, how things were at home. I talked to her about school, Hazrat about college. We were in such a rush of excitement we all gabbled over each other. It was only after we'd put the phone down I realized we hadn't really talked about the journey itself. I'm not sure I could have even begun to find the words to tell her what I'd been through. Some things are better left unsaid.

Just hearing her voice changed everything for me. I felt as though I had a renewed sense of purpose. She'd sacrificed so much to get us here. I needed to make her proud.

Starting Point used my second suicide attempt as the catalyst to really go into battle for me over my age dispute. For months they had been observing me in the classroom, carrying out their own age assessment report on me. They told both Kent Social Services and the Home Office that, in their view, I was a child of fourteen.

I was by now Starting Point's longest-serving pupil. Most kids came there for about six weeks, until they got a place at a local school. The whole idea of the center was to immerse children in education and English language before they were placed in mainstream schooling. But until I could prove I was fourteen, no school in the area could take me.

"You will stay here with us until we work this out, Gulwali," Mrs. Kellett told me. "If you stay two years then so be it, but we will not let you down."

She and Mrs. Brodie were the only two adults I believed and trusted.

Thanks to their interventions I was able to secure a new interview to discuss my claim for asylum and to appeal the decision to reject me. Nassi, my social worker from Kent, was there, but the interview was carried out by Bolton Social Services.

The meeting was a similar format to the ones I had attended before: five officials sitting around a table with a translator to ask me questions. But this time it could not have felt more different. This time I had people on my side. Nassi and Mrs. Kellett, as well as an advocate from Action for Children, a national charity dedicated to helping vulnerable and neglected children and young adults like me.

After the meeting, Nassi filed a new report in Kent. By now, she had my birth certificate and Afghan passport. Kent spoke to Starting Point, saying they were prepared to concede I was younger than they had previously documented, but that they wanted me to accept I was born in 1992 not 1994, making me just under sixteen. It was crazy. How could I accept it? That was not my true age. But, instead of getting upset, I found it funny. This new offer meant things were at least beginning to move in the right direction.

Finally, after a lot of back and forth, Kent Social Services officially reassessed my age, and accepted what I had told them from the beginning: that my birthday was October 11, 1994. The Home Office also then wrote to me saying that they would reconsider my case. I was given discretionary leave to remain in the UK until just before my eighteenth birthday.

It was the happiest day of my life since I had left home. And not long after that, Hazrat was allowed to become my legal guardian, and he and I were allowed to rent a house together in Bolton.

STARTING POINT STARTED TO TRY TO FIND ME A SCHOOL. I WAS sad to leave the place that had become my sanctuary, but I was delighted to finally be allowed to go to a proper school. I was interviewed by Essa Academy in Bolton, by Mr. Khaliq.

"Do you think you can cope with this, Gulwali?" he asked. I already knew him because I had seen him at Starting Point. He had always used to joke with me that I'd come to his school one day but that I wasn't to expect treats like the tea and toast Mrs. Brodie made for me.

"Yes, I do."

There were only two months left of the academic year. I asked them to put me into year 9, the academic year below my actual age group so I could try to use that time to catch up.

When year 10 began, all of the other pupils were choosing their GCSE subjects. I was very frustrated to be told my English wasn't good enough for me to be allowed to sit any exams; instead, I was placed in entry-level groups in math, science, English, and other basic subjects. I was also working toward a certificate in care and social work. The school said that without any other qualifications, that would be one of the only future career routes open to me.

The hardest thing about starting mainstream school was making friends. I had gone from a center with a maximum of fifty pupils to a school with one thousand. The other children scared me. I couldn't understand why some of them didn't want to learn and messed about instead, being rude to the teachers. I also hated sports and especially loathed getting changed in front of everyone during PE. That was a culturally difficult thing for me to do, and I cringed red with embarrassment as I put on my PE kit, the other boys laughing at my discomfort.

One of my favorite moments was a school geography trip to Scafell Pike, the highest mountain in England. We climbed it. As the other children huffed and puffed and moaned, I strode up like a champion. After so many mountain crosses and treks on my journey, the modest ascent was a breeze.

CHAPTER 32

SLOWLY BUT SURELY, I SETTLED IN AND MADE FRIENDS. I was still living with Hazrat. Now that I was officially a child in the eyes of the state, social services kept a much closer eye on me, referring me for more counseling. The counselor was really nice but it still felt culturally weird for me to tell my problems to a stranger. I was much more comfortable talking to Mrs. Brodie and Mrs. Kellett, adults I had come to love and trust, and who I finally realized were not going to reject me however angry or confused I was.

I threw myself into school life and activities, doing my Duke of Edinburgh bronze and silver certificates, a brilliant personal development program for young people founded by the Duke of Edinburgh in the 1950s. It made me laugh to think that here I was, an Afghan, on a course designed by the queen's husband. Once the very thought of such a thing might have felt like a huge betrayal of everything I stood for. Now I saw participating in it as a genuine privilege. The program was really tough and covered all sorts of things from learning a new language to sports and volunteering. Part of it involved water-based activities, such as kayaking and life-saving skills, so I had to learn to swim.

I was shaking like a leaf when I got into the water for my first swimming lesson. I honestly did not think I could do it, not after my experience of nearly drowning in the Aegean Sea. But the other young people on the course and the instructor encouraged me as I stood quivering by the edge of the pool.

"Come on, Gulwali. We'll look after you. You can do it."

The instructor held my hand tightly as I gingerly allowed myself to be led into the shallow end. That I did it was a massive turning point for me. If I could do this, I could achieve anything.

Six months into school year 10, I knew that my English had vastly improved. I went to see the head teacher, determined to convince him to let me sit my GCSE exams.

"Okay, Gulwali, you can take them but don't be disappointed if you fail them."

I chose IT, Urdu, religious studies, and geography, as well as the core subjects of English, math, and science.

Religious studies was strange but fascinating. It was the first time I'd learned about Christianity and Judaism. What I couldn't get my head around was why people of different religions were so often at war with each other, because what I learned is that all the Abrahamic religions are based on the same essence of one God.

I was more freaked out by the citizenship and social studies classes all pupils had to take. In these, we discussed things like moral ethics and sex education. I couldn't understand why anyone would want young people to know about sex.

In my head, telling teenagers how to get contraception would only encourage them to do it. We would never, ever discuss anything like this in Afghan society, not until you are married. It was all I could do to not run from the classroom with my hands over my ears.

I had come so far by now but I wasn't quite there yet. I had stayed in touch with my friend Mrs. Brodie from Starting Point; I visited her once a week, and she cooked me dinner. Her family—her husband and two sons—were lovely to me and I loved going there to see them all. One son was a keen rugby player. I told him I wanted to learn how to play it but that I didn't want to get hurt.

"Gulwali, I love your enthusiasm, but if you don't want to get hurt and you hate changing in front of the other boys"—he laughed—"I don't think rugby is the sport for you."

In the second year of my GCSEs I started struggling again. The subjects were hard and my depression was eating away inside of me. I knew that if I let myself slip I'd be suicidal again. I decided to try to help myself stay positive by doing something positive for others.

The school appointed me the school ambassador for international arrivals. I had been one of those lost, scared children from a foreign land, and I knew how hard it was to fit in. I could help them. With the help of my teachers, we devised a school plan to give extra moral and academic support to the arrivals, ensuring that immigrant pupils would swim not sink when they got there.

The plan worked so well that other schools in the area copied it. That made me so proud. I also became a member of the school council and pupil volunteer librarian. For me that was a joy: I would much rather spend my free time in a library than anywhere else.

IN OCTOBER 2010, WHEN I TURNED SIXTEEN, HAZRAT HAD TO go back to Afghanistan in order to be with my mother and siblings. This was a very difficult decision. He knew he was heading back into danger but there had been some serious incidences that led us to believe our mum and siblings weren't safe. I was terrified I'd never see him again but I tried to be brave. He and I both felt that she'd risked all to save our lives. Now her life was in danger and it was his turn to be there for her. Saying good-bye to him and being left all alone again was incredibly hard, but I coped with it because I was relieved he would be there for the family.

Social services were still officially responsible for me, and they wanted me to either return to Kent or be placed in foster care.

I had been really desperate to be in foster care the first year I had arrived in Britain, but now I was a star pupil with friends. I had learned the hard way to cope with living on my own perfectly well, yet only now I had got used to it, they were insisting I couldn't. As ever, the system seemed illogical.

In December 2010, Mrs. Brodie drove me to meet a potential foster family. Their house was a large, former vicarage in the countryside near Bolton.

"It's the middle of nowhere," I complained. "Let's just not bother going."

As gently encouraging as ever, Mrs. Brodie insisted we at least go meet the family before I made my decision.

It was snowing heavily on the day she drove me there. I had by now been living by myself for two months, without Hazrat and with no central heating. I think it was only the cold that made me agree to go and look at a foster family.

They were a married couple called Sean and Karen. They had a son of their own, who was grown up and lived away from home. They also had two other foster children, a boy and a girl, both British.

As Sean showed me around the house and where I might sleep, I was polite but firm: "Thank you, but I am not coming."

I was still insisting that I be allowed to live on my own. But Mrs. Brodie knew me well enough by now to know I was my own worst enemy at times. She badgered away at me until she convinced me to give it a try.

Just before Christmas, I moved in. What I couldn't get my head around was that Sean was cooking the dinner, not Karen. I had changed so much and had had my eyes opened to the world, but in some places, deep inside, I was still the same conservative village boy I had always been. I struggled to understand why any married man would cook when he had a wife. I thought this made Sean less of a man and, rudely, I didn't want to eat his food.

But that night, as he made sure I settled in, I was touched to see they had bought me a desk for my room. It was old, made of solid oak. I ran my fingers over it, thinking what a nice gesture it was.

The next night I ate Sean's food and told him it was delicious. I meant it. He offered to give me cooking lessons, which I accepted. Over the coming weeks he showed me how to cook curries, as well as Italian and

Mexican food. He explained to me that he did all the cooking because he loved it and because Karen, or Auntie Karen as I called her, worked very long hours. He said marriage was about partnership and about being a team. As I observed Sean and Karen, who seemed very happy and in love, I began to think differently. I began to see that my views were borne of a different culture, a culture I still loved and believed passionately in, but that it was not the way people lived here.

I think I realized that I could respect this different culture, while still honoring my own ideals and beliefs. No one here was forcing me to reject Islam, to become secular or a Christian. They accepted me for who I was so I had no reason to reject them. Living with a foster family was like a cultural fast-tracking, and possibly the single biggest thing that helped me truly begin to assimilate in Britain.

I continued studying hard, and Sean offered to help me. One night, not long after I moved in, I handed him a piece of paper with a timetable on it: "Here you are. This is my homework schedule."

"This is for you, right? You want me to help you stick to it?"

"No, it's for you. These are the times I need you to help me with my homework."

He laughed so hard I thought he might fall off his chair. "Gulwali, this is a first. I've never had a foster kid give *me* my homework schedule before."

Living with Sean and Karen also really helped the depression to fade away. I was doing well at school and loved coming home to share my day and show them my positive reports. I'd collected so many certificates and badges of achievement by now that Sean used to tease me, saying all the badges on my blazer made me look like a general. I had love and warmth, and I basked in the glow of it.

By now, I was in year 11, and about to sit my GCSEs. I was made an ambassador for the whole of Essa Academy, a prefect, a school counselor—I threw myself into school life and I felt that I really belonged.

I surprised my teachers with my GCSE results: I got an A in Urdu, and Bs and Cs in the other subjects. My head teacher, Mr. Badat, couldn't believe that I had managed a C in GCSE English; I was the only person in all of my core subject classes to get a C or above in everything. When I received a B in math, having covered the whole syllabus in six weeks, the head of math, Mr. Hussein, named our year "Gulwali's Year."

In September 2011, I moved to Bolton College to take my "A levels" in politics, economics, philosophy, and Urdu. I also undertook an extended research project in the issues young refugees face around education.

I pursued my political activism outside school, too, becoming a regional advocate for children in care, mainly refugees. I stood for the UK youth parliament election. I lost but, as a result of that, I was recruited by the British Youth Council and asked to join the National Scrutiny Group, which had been set up to advise the government on policy and how it affected young people. I was one of only fifteen young people nationally on the panel. I also joined the youth wing of the Labour party. I wanted to learn more about the British political system because I was really interested in being an activist. I felt the Labour party was the most diverse, so it would be where I would fit best. But of course I still had huge issues with Tony Blair and his decision to take part in the invasion of my country and that of Iraq. But I knew that not all the Labour party members shared his view.

I do know I was still very confused though. At times, rage still overwhelmed me. For all of my progress there were some days I felt like I was going backward, consumed with guilt, confusion, and frustration at the alien nation I was struggling so hard to live in. When I was around other young Afghan refugees, those feelings intensified; all of us felt dislocated and full of loss. Talking about politics or the problems back home only made it worse.

Some of my friends got involved in extremist radical Islamist groups. These groups prey on the vulnerable and lonely. They offer friendship and brotherhood and are masters at seizing on and manipulating a person's traumas or unresolved issues. I was still angry inside and struggling to find my way in a new culture, so it wouldn't have been hard to brainwash me. I was invited to attend talks and certain mosques that definitely had more radical agendas, but luckily for me I was too busy to go because I had all my other activities going on. Without that, I honestly admit, I might have gone down the extremist path myself. But over time those solid connections through school to other youth organizations saved me. I was learning that I had the opportunity to be heard and to influence policy and decision-making processes. I knew these were brilliant chances I had been offered, and to me it felt like something far more useful and productive for me to be doing than being drawn into destructive groups that only promoted hatred.

I was also so busy trying to improve my studies and learn English that I didn't have time to engage with people who only wanted to complain. But I can see why, in the absence of any alternatives, some of my acquaintances got dragged into that. They were lonely, and had nothing to do, no money, and no support. Isolation drove them into the hands of radical preachers.

I was also becoming hugely ambitious. I had worked out that the better I did and the more I achieved, the more chance I had of one day returning safely home. I began to think that if I could go home, I could help to make Afghanistan safer, so that no child had to experience what I had. In a nutshell I think I was beginning to challenge my rigid mindset and develop new ideas whilst reaffirming certain long-held beliefs. I really began to appreciate that Great Britain is a place where people from all over the world can live in peace. A place of freedom where you can make things happen.

I honestly don't know how or why I was so driven at that point. It was also my way of giving back to British society. I had been given so much from the UK, I wanted to do something in return for my adopted country. And I also wanted to help other young people. I knew what it was like to have no one. After all my awful experiences, I didn't want to waste a single second. I wanted to suck up every opportunity, every chance, every experience that came my way.

On my weekends I attended an access course at Manchester University. It was designed to help get young people from disadvantaged backgrounds into top universities. If I passed that, it would add to my university entrance score. I was worried I might not get the grades I needed to follow my dream of studying politics, so this was the way of ensuring I got the place on the course I wanted.

IN OCTOBER 2012, I TURNED EIGHTEEN. MY LEAVE TO REMAIN had expired and the Home Office was once again considering my asylum application. It was a very tense and worrying time, especially with my "A levels"—university entrance—exams coming up. I was determined not to fail. With Sean and Mrs. Brodie's help, I wrote letters far and wide to anyone I thought could help influence my case, from members of parliament I'd met through the Labour party, organizations I'd been involved with, teachers, and other people who knew me. My message to the Home Office was this—Afghanistan is not safe for me. Plus, I've worked really hard, I'll continue to work hard, and I will contribute back to this country if you let me stay. But sadly, asylum cases aren't decided on how hard someone works and what they give back. I knew that, but I tried anyway. In all probability I expected the Home Office to say no. The stress of the case was keeping me awake at night and I really worried I was going to flunk my exams.

Now that I was an adult in the eyes of the state, it was time to move on from Sean and Karen's. I was given independent accommodation in Bolton town center.

I had spent two years with Sean and Karen and didn't want to leave; it had been the happiest time in my life since early childhood. But they remained in very supportive close contact and encouraged me to focus on my studies, despite the uncertainty of not knowing whether I faced deportation or not.

In the end, my "A level" grades were better than expected: I was thrilled with one A and two Bs. I also got an A star for my extended-study project on refugees, and was awarded a sixth-form excellence award. Not bad for a boy who just four years earlier had spoken only a few words of English and had come to England in the back of a banana truck.

And then, finally, came the news I had been waiting for: I had won my appeal. I had been granted asylum. I wept with joy as, hands trembling, I read the letter over and over again.

WE HAVE A SAYING IN PASHTU: "*PA JAMOO KAI NA ZAYDAM*." IT roughly translates as "Feeling too big for our clothes due to pride."

A full five years after that boat nearly capsized, I stood in Burnley town hall with my foster parents, Karen and Sean, by my side. I had clean, fresh clothes now, and I was definitely feeling too big for them. I, the once-scrawny refugee, had been selected to carry the 2012 Olympic torch through Britain ahead of the London Olympic Games.

Mrs. Brodie, whom by now I was so close to I called "Mummy number two," had suggested I apply to do it: "Well, what have we got to lose, Gulwali?" I loved how she always referred to me as "we." It made me feel special, loved.

When I learned I had been selected, I was so proud I thought I might burst, but I also felt incredibly humbled.

The day had begun early, with a forty-five-minute drive from Sean and Karen's home in Bolton, to meet the rest of the Olympic relay team. The streets were lined with people cheering and waving flags. At that moment, I don't think I could have loved my adopted country more.

As I set off, Karen warned me to walk slowly and savor the moment. Police were on all sides of me, the crowd ecstatic. I tried to walk slowly but in my excitement I was bouncing along. I kissed my torch, beaming with pride, with love, with recognition that this life-changing moment was one of the most special things I would ever do. At so many times on my journey to freedom I had felt hopeless, despondent, and afraid. Many times I considered giving up and going home. But at those moments of weakness one thought had kept me going: my mother sent me away to save my life.

My mother sent me away so she didn't have to bury another person that she loved. In doing so, she had made the ultimate sacrifice any mother could ever make.

As I ran through the streets of my adopted second home, the torch burning brightly, with people cheering and taking photos, I thought only of one thing—her.

At the moment I knew, beyond all doubt, that I hadn't failed her.

CHAPTER 33

I CAME BACK TO EARTH WITH A BUMP THE NEXT DAY. THE local newspaper had run a story about me carrying the Olympic torch.

I logged onto my PC excitedly to look at the piece online.

There was my picture and some really nice words about how much it had meant to me and what a special event it was for the country. Then I noticed the comments. There were loads of them. Really nasty, racist, virulent comments, asking what right I had to be doing such a thing and why didn't I go back to my own country? I had to laugh at that. I'd love nothing more than to go back if I could. Others said it was disgrace this honor had been given to me when more deserving English people should have had the chance. I agreed with that—many people deserved it too—but I had applied and been selected. How did that make me so bad? I had been so proud to have a tiny part in something as vast and wondrous as the Olympics, an event that transcends race and politics. Those comments deeply hurt. It was my first real taste of racism. I couldn't get it out of my head.

Some days I woke up thinking that if people here hated me so much why didn't I just hate them back? Wasn't it them who had bombed my country after all? Wasn't it their fault I had had my childhood destroyed and had walked halfway around the world? Every time I looked in the mirror and saw my still-scarred face I remembered Calais and the awful dehumanizing horror of it for the people still trapped there. I felt so guilty that I had escaped and other people were still stuck there. My nightmares took a turn for the worse.

Other days I just wanted to lie in bed and not get up. I missed my family so much, and without Hazrat life was so much harder on every level. To make matters worse Qamandin was facing deportation too. To top it off I also faced something of a backlash from a small minority of other Muslims who accused me of selling out by carrying the torch. I felt like however hard I tried I just couldn't win.

Anger was my obvious reaction. But it was a lost, stupid, helpless anger. I just needed to be understood, to be heard and to be healed.

But, just as on my journey, when things got really tough I found solace in prayer and contemplation. God hadn't let me down this far and he didn't now.

I enjoyed talking about religious ideals with other Muslims but also friends of other faiths. Through my youth activities I now had friends who were Hindu, Christian, and Jewish. I respected our shared values and traditions, but talking to them also helped to reinforce my own faith. I suppose it was a similar process to the one I had already been through culturally, learning to give and bend but also accepting and loving who I was and what I believed in.

My head never stopped swirling and I wondered why it was I thought so much. But I also think my constant brain activity was a kind of defense mechanism. The moment I stopped being busy or thinking about how to improve myself, I started to get depressed and angry again.

In 2013 I started university. All the hard work paid off and I won my place at Manchester to study politics. Walking through the grounds on my first day was quite possibly the happiest day of my life. I just couldn't stop smiling. I was nervous—there were so many students there from all sorts of backgrounds. I was still a bit freaked out by seeing female students in short dresses: old habits die hard. But I knew they had every bit as much right to be there as I did. And they'd worked just as hard to win their places. All the freshmen were as one that day.

Of course I didn't do a lot of the traditional freshman-week things like getting drunk or going to crazy parties. But I went to lots of fringe events and still had loads of fun and made friends quickly. I felt like I'd found my rhythm among the lecture halls, books, and feverish desire to learn. I was getting the bus home on my third day, my head resting against the window in tired satisfaction, when something very strange happened. I heard my maternal grandfather's voice again. The same voice I had heard the day I crossed the border from Iran to Turkey the second time, at a time when I felt I couldn't go on. "Life is an education, Gulwali." I swear I heard this as if he were standing right next to me.

Yet, once again, the moment I felt settled, the emotional rug was pulled from under my feet. I woke up in my fourth week unable to get out of bed. I had had one of my worst-ever nightmares. When morning came, I just couldn't lift my head from the pillow. I felt choked, panicked. The idea of even going outside, let alone to university, filled me with terror. I stayed in bed all day, as the clock dragged by in misery-filled minutes that felt like hours. It was one of the worst and scariest days of my life. I felt hate, bitterness, and anger floating through the room as clearly as if they were physical presences. I heard them whisper my name, their tentacles stirring in my soul, willing me to succumb to their poison. A lightless sky forever falls on parts of everyone's journey, regardless of where they are. It would have been so easy for me to give up, perhaps attend an extremist meeting that night, go onto social media and rant about the injustices of it all, or even try to kill myself again. If I had had enough pills next to me, I think I might have.

I lay there with the curtains closed, fading slowly into the surrounding blackness, trying to concentrate only on my breathing; trying, in vain, to quiet the voices at war in my head. I turned over. Slivers of sunlight poked through the top of the window and with them came a sudden kind of comfort, a slow-dawning realization: *jihad*.

Slowly, I took out my Quran, turning through the pages reverentially, just as my father and uncles had taught me, searching them desperately for the absolute word of God, written for the day of judgment. I turned to the relevant pages.

> You shall spend in the cause of God; do not throw yourselves with your own hands into destruction. You shall be charitable; God loves the charitable. (2.195)

> You shall resort to pardon, advocate tolerance, and disregard the ignorant. (7.199)

For the first time I truly understood what it meant. Not the most manipulated, twisted concepts of *jihad* as "holy war." That's the false version used by terrorists acting in the name of Islam to commit terrorist acts, aimed at the indiscriminate killing of innocent people. The literal meaning of *jihad* is "struggle" or "effort"—the holy war within oneself. That's the battle I had been warring with myself all this time, the confusion and pain that had been crippling me since I fled Afghanistan. And, I understand now, this is the battle I would have to continue to fight. *Jihad* is a battle within all of us: it is a fight we all must fight in different ways—whatever faith we may come from or if we have no faith at all. That day, as the sun continued to fill my room, I knew beyond all doubt that I need my *jihad* so that I can go on loving. I fight *jihad* to be.

EPILOGUE

THERE IS ANOTHER PASHTU SAYING: "THERE IS NOT enough time in this life for love—I wonder how people find time for hate."

I could all too easily have lived a life of hate. I was twelve years old when my father and grandfather were massacred. All I cared about, loved, respected, and was influenced by was torn away from me in a single instant.

It would have been easier for me to choose anger. It's in my genes. As a Pashtu, the notions of honor and family are at the heart of my identity. We are attached to the concept of revenge and blood feuds, which can go on for generations.

And then other men, soldiers from both sides of a conflict I was too young to understand, wanted to use my brother Hazrat and me as pawns in their game of war. Whichever side we would have chosen to work with, the result would have been the same—in all probability we'd be dead, and more people would have died with us. My mother would have had to bury more of her loved ones, more families would have wailed in grief, and my country would be no further along the road to peace.

Violence begets violence.

The choice to go away was not mine; it was my mother's. She proved her wisdom beyond doubt when she sent my brother and me to Europe. She knew that, were we to stay, the outcome would have been bloody and ugly—not a thing of beauty, honor, or justice. Yet, so many times, in my desperate and loneliest moments, I was left bewildered, despairing, and angry at her decision, my childhood a brutal game of survival. So many times on that awful journey I nearly didn't make it—jumping from the speeding train; coming so close to drowning in Greece; on those endless treks without food or water when my young, exhausted body wanted to give up and fade into blackness.

In part, I was saved by the warmth and kindness of the friends I made who looked out for me. More than anything, this book is about faith, hope, and optimism. I hope too that it is about dedication and commitment toward fellow human beings. A story of kindness, love, humanity, and brotherhood.

Baryalai, the kind and lovely man who truly took this lost child under his protective wing, did make it to the UK. We found each other again in 2008 and we met up a few times. We lost contact about three years ago. He simply disappeared, and I fear he may have been deported or gone into hiding. His is one of the few names in the book that remains unchanged because I hope he will somehow read this, and in doing so, we may meet again.

Mehran, the constant joker who kept me going with his humor, lives in Greater Manchester and works in hospitality. He's doing just fine.

Sadly, I haven't seen or heard from Abdul, the fourth member of my original group, since the day I escaped from the police prison bus in Iran. I don't know if he made it to safety or not.

Nor have I seen Shah or Faizal since the day they were put into a separate vehicle from us in Turkey.

Tamim reached the UK in 2009 but I last saw him in 2012, and I don't know how he is now. I fear he may also have been deported.

Jawad I haven't seen since he was left behind in the *musafir khanna* in Istanbul, but we are in contact by email. He made it as far as Greece, where he now lives and works.

Hamid, ever the smart one, lives in London. He is also at university, studying to be a doctor, and I know he will go on to great achievements one day. He and his friend Ahmad arrived in the UK at the same time, having managed to stay together since I got into the back of a truck in Greece and they did not. Ahmad also lives in London and is awaiting the final Home Office decision on whether he will be allowed to stay in the UK.

Jan, Qamandin, and Engineer also made it here. I last saw Engineer in Manchester, in 2009. He had done okay and was studying at college, but then the Home Office refused his claim for asylum and he left Britain.

Qamandin currently lives in Kent and also continues to fight the Home Office for the right to stay here.

Shafique arrived in the UK one month after me and we were reunited in the Appledore immigration center for unaccompanied children. We are still very close friends but, as yet, he does not have a final decision from the Home Office, and I am trying to help support him with that.

Hazrat returned from Afghanistan in 2014, after three years back at home helping my family. Now his case is back with the Home Office.

I call my mother once a week. But perhaps the hardest thing of all I have been through is that when we talk, we have little so understanding of each other. I try to talk to her about my political life, my university exams, and my campaigning, but my life here in Britain is like a different planet to our old life in Afghanistan. By sending me away, she definitely saved her son, but she also lost him. Of everyone, my mother paid the heaviest price. But she will always remain my inspiration, as will my beloved grandmother, who passed away a couple of years ago. We were able to speak on the phone when she was dying. She told me she loved me, and I her. I ached to be able to hold her and stroke her hair as she breathed her last.

Hopefully the opportunities and education I have been given here will allow me to get a job with something like the United Nations.

When I graduate from university, that's my goal. After that I hope to begin a slow reverse journey back home to work for my own people. My ultimate dream would be to become a politician, or maybe even the president, of Afghanistan.

I truly believe faith and fate brought me on my journey, and I want to thank Great Britain from the bottom of my heart for taking me in. I hope I have given back.

But at no point did I ever think I would be staying in the UK forever. Home will always be home. And that's Afghanistan.

But these days I know what I truly am proud to be. And that is a global citizen.

Through my political activism and community volunteering I hope I have helped to improve other young people's lives and helped the public understand the plight of refugees and asylum seekers. Recently my old school launched an annual pupil award called "the Gulwali Passarlay Award for Overcoming Adversity." That I can inspire another young person to work hard and overcome is amazing.

Ultimately, that is why I wanted to write this book.

But there are still some days, even now, when I wake up after a particularly bad nightmare and it's all I can do to get through the day. The nightmares will never leave me, nor will the memories of those I have lost, nor the faces of the terrified and vulnerable refugees I met on the road. Did the tiny little girl in the blue bobble hat I saw on the mountain crossing into Turkey ever reach safety? I have no idea who she was or from what war-torn country her family had fled, but her tear-stained, grubby face and scared eyes are forever seared into my soul.

I can't and won't ever turn away from her memory. The enemy of love is not hate, it is indifference. The enemy of love is turning away from those in need. The enemy of love is doing nothing when you can help your fellow man.

The refugee crisis—the greatest global crisis since World War II—has been caused by conflict, wars, poverty, injustice, and oppression. It is our moral duty to treat these fleeing human beings with dignity and respect. We cannot shy away from the fact that recent wars in Iraq, Libya, Syria, and Afghanistan have caused this crisis. Nor can we pretend that the Western desire to buy cheap products or possess the latest must-have items at a bargain price does not contribute to poverty and inequality.

True freedom and democracy is a two-way street. It demands that people educate themselves about the world around them. That requires an honest and inquisitive mind—one that questions all opinions, yet hates none. If a person wishes to be free, then they must understand the shackles that bind them. The Internet and the age of social media make the dissemination of ideas very easy. There is no way we can turn back the clock on this reality; instead, we must learn to live with it. It can be empowering, or it can be destructive. Causes are powerful—and now they travel vast distances instantly.

When immigrants and refugees don't feel welcomed or part of society, that's when they can turn against it. For many young Muslims it feels deeply unfair to be constantly asked to apologize for terrorism or to be blamed for it when it's nothing to do with them or their lives. Look at the example of Ahmed, the little boy who recently took a clock to school in Texas. All he wanted to do was proudly show it to his teacher. But instead he was suspected of bomb making and led away in handcuffs. What I loved though was how Americans of all ages and cultures came to his support. His father said this is "what made America."

America has citizens of all walks of life who love it; that the American dream is still possible is why so many immigrants still flock there. But sadly it does also have extremists in its midst. But to tackle them we need to understand them.

Many extremists are drawn by the romantic notion of fighting, and maybe even dying, for a righteous cause. Perhaps that's why young people from every walk of life are at risk of radicalization: whether they are a frustrated white male who walks into a black congregation and starts shooting at innocent people, a desperate Somali fisherman's son who opens fire in a Kenyan mall, or a French schoolgirl born of refugee parents, who resents the strictures of life and seeks adventure and fulfillment in what she believes is the exciting new world offered by the false prophets of the so-called Islamic State—a group of people who in no way represent Islam.

Assumed thinking creates assumed responses. Our world is a fast-changing place—but more and more people are becoming entrenched in the same mind-set as previous generations. I don't see how this can be helpful.

My journey nearly killed me, and it left mental, physical, and emotional scars that I will bear for the rest of my life. Those moments that I was at my lowest, when I felt as though I could not keep going—those truly painful moments where death seemed the only solution—those moments almost entirely coincided with periods where I struggled to move forward and to progress my journey. And so it is in life. We are happiest when moving forward.

I have had the help of many great friends on my journey to help me continue to do so. And I still do to this day. None of us travel alone in life. We all have the power to help those around us, or to harm them. It is the choices we make that define our walk, define our own personal journeys, and make us the people we are.

I want to thank you for reading my book and for being a part of my journey. Truly. Thank you. But please don't change the world for me. Do it for the other children out there alone in the world, lost, afraid, and trying to find safety. Do it for the mothers who would rather send their children into the unknown than to see them die of starvation or from bombs raining from the sky. Do it so that no other child has to wake up to another lightless sky in the way I did.

If I have one single dream it is this: that a child in the future will read this book and ask, "What was a refugee?"

We can change the world. All of us together. We can.

We can end this.

Co-Author's Note

Making his journey over the course of a year, and through eight countries, Gulwali saw more and suffered more than any twelve-year-old child should.

He is not alone.

In 2014, there were 23,100 asylum applications made in the twenty-eight European Union member states by unaccompanied minors (defined as persons under the age of eighteen who enter without an adult, be it a parent or guardian). Today, more than half of the world's refugees are children.

As we wrote this book, in the summer of 2015, a UNHCR Global Trends report, "World at War," revealed shocking new statistics that showed that the worldwide displacement of people is at its highest level ever recorded. By the end of 2014, 59.5 million men, women, and children had been forcibly displaced by wars, conflict, or persecution.

If this number of people was a country, it would be the world's twenty-fourth biggest nation.

Just a decade earlier, that figure was 37.5 million. The staggering rise is due to several factors, including recent conflicts, political repression, and natural disasters. Some experts predict that climate change alone could create as many as 150 million so-called environmental refugees by the year 2050.

The reasons why people flee the countries they come from may shift and shape at any given time. Currently, Syria has the world's highest number of refugees; at the time Gulwali made his journey it was Afghanistan. What does not alter is the way in which smugglers operate. It is perhaps simplest to describe the business of smuggling by looking at its structures as though it is a corporation:

At the top level are powerful yet rarely seen national agents, the "CEOs," men who control smuggling in certain countries. They use a variety of aliases. Gulwali only knew his as Qubat. His family only met Qubat once and paid $8,000—the amount agreed to get Gulwali as far as Italy. Money was lodged with a third party agreed upon by both sides, and only after Gulwali made it across each border was the next tranche of money paid to Qubat. Agents at this level build repeat business by reputation: it is not in their interests to cheat or to fail. But it must be pointed out that had Gulwali died along the way, no refund would have been given, and payment would still have been required to be made to Qubat in full. I have in the past interviewed many families with dead children for whom this was the case.

Underneath this are the regional agents—the men who control a certain area of a country, akin to senior management level. Malik, the be-suited and briefcase-carrying man Gulwali met in Turkey would fall under this category. Men like Malik, who also owned the brothel where Gulwali briefly stayed, are very often where the cross-over between smuggling and trafficking collides, profiteering from both.

Refugee smuggling—reportedly a $7 billion global industry today—is defined by the UN and Amnesty International as "procurement for financial or other material benefit of illegal entry of a person into a state of which that person is not a national or resident." On the other hand, human trafficking is characterized by exploiting another human being, moving them by force and against their will. The UN definition of human trafficking includes the recruiting, transporting, or harboring of people by means of force.

Yet as Gulwali would like to point out, this definition can be a murky one. If a refugee in the hands of paid smugglers is forced into taking a dangerous route against their will, as Gulwali almost was when being sent overland to Greece, then where does the line between smuggling and trafficking fall? There have been several documented cases of smugglers forcing refugees at gunpoint to board unseaworthy ships, throwing refugees overboard, or locking them below deck, thus condemning them to certain drowning. Is this then smuggling or trafficking?

The next level on the rung is effectively middle management: the guesthouse owners or men who can offer specialist logistical solutions, such as a fleet of cars. Black Wolf falls into this category.

Underneath that is the operational level, the myriad networks of smugglers, boat captains, and drivers, who are paid to move people from place to place. This is where things become increasingly unstable and most often go wrong, and where refugees can be at their most vulnerable.

At the very bottom of the smuggling business are people like Serbest, the man who provided Gulwali a horse and guided him over the border into Turkey for the second time. Gulwali has a great deal of sympathy for many of the people he met who worked at this level, because so often they were fighting to survive in situations of great poverty and conflict in a way not dissimilar to himself.

Names and certain details throughout the book have been changed to protect identities and certain characters have been omitted. We have sought to be as accurate as possible, but it must be stated that these are the sometimes hazy memories of a twelve-year-old child; dates and times blurred into one on the road. Gulwali had a sense of the months based on the weather, but his journey took him from high mountain passes to stifling basements, making it hard for him to tell at times.

What are described as detention centers at times in the book may have been prisons, and vice versa. Often it simply wasn't clear to him where he was being kept, only that he was incarcerated.

Returning to the detail was both extremely painful and traumatic for the author. Gulwali, like most Afghans, is a natural linguist and picks up phrases quickly. As he says: "I had to, in order to survive."

His English developed quickly because it is the common language used across Europe by many police and officials in their attempts to communicate with refugees. However, at certain points within the book, we have taken small liberties with his understanding of language in order to maintain the narrative.

We simply could not include every story, every voice, or every person he met along the way, as much as we would have liked to. There were many characters who sadly remain unheard or unwritten. Above all else, Gulwali and, I hope, this book will give voice and a human face to the refugee crisis. The very fact this book exists ensures that these people have not been forgotten.

At the time of writing—the end of July 2015—over 2,000 men, women, and children are known to have drowned in the Mediterranean this year alone. The unofficial figure may be far higher. Their story ended in the way that Gulwali's book begins. Gulwali could so easily have been one of them, his voice snuffed out in the depths of a cold sea.

I leave you with that thought.

—Nadene Ghouri, August 2015

Acknowledgments

I would like to thank everyone who has helped me along my life journey. But special thanks go to:

All those I traveled with who helped to keep me safe. My social worker, Nassi; and all the teachers at Starting Point, but especially Katy Kellett, for believing in me and fighting for me; and Chris Brodie, my mentor, best friend, and second mother. Thanks to all the staff at Essa Academy, in particular Mr. Badat, Mrs. Reid, and Mrs. Grills for helping me find my political and campaigning voice. Also Mrs. Bolton, who I always went to whenever I felt down and upset. Thanks to everyone at Bolton Sixth Form College, particularly Mr. Hindle and Mr. Ivory, who both taught and encouraged me. And to all at the University of Manchester, especially Dr. Julian Skyrme. To the various youth workers and organizers I have met through different committees, commissions, forums and groups—you are amazing.

I am forever indebted to my foster parents, Sean and Karen, for their warmth, love, and support, and for allowing me to share their home and family.

To the inspirational Nadene Ghouri, without whom I wouldn't have written this book. Her passion and enthusiasm kept me going to the end, helping me through the stress and pain. Thanks also to her husband, Sam Robertson, and everyone at Gladstone's Library for nurturing us during the writing process.

Finally, thanks to my lovely agent, Brandi Bowles, and everyone at HarperOne for believing in this book and making it a reality.

There are so many others who have been a part of my journey. I am sorry if your name isn't here, but I hope you know who you are and how important you are to me. I am in debt to so very many people.

THE LIGHTLESS SKY

A Twelve-Year-Old Refugee's Extraordinary Journey Across Half the World

GULWALI PASSARLAY

with NADENE GHOURI

HarperOne
An Imprint of HarperCollins*Publishers*

HarperCollins books may be purchased for educational, business, or sales promotional use. For information please e-mail the Special Markets Department at SPsales@harpercollins.com.

HarperCollins website: http://www.harpercollins.com

FIRST HARPERCOLLINS PAPERBACK EDITION PUBLISHED IN 2017

Designed by Ralph Fowler / rlfdesign
Map, pages iv-v, by Jamie Whyte; used by permission.
All photos courtesy of the author, except page 3 of the insert, courtesy of Naveed Yousafzai, and page 8 of the insert, courtesy of Capture the Event.

Library of Congress Cataloging-in-Publication Data is available upon request.

ISBN 978-0-06-244389-2

17 18 19 20 21 RRD(H) 10 9 8 7 6 5 4 3 2 1